UNBELIEVABLE

UNBELIEVABLE

MY FRONT-ROW SEAT
TO THE CRAZIEST CAMPAIGN
IN AMERICAN HISTORY

KATY TUR

DEY ST.
An Imprint of WILLIAM MORROW

HarperCollins books may be purchased for educational, business, or sales promotional use. For information, please e-mail the Special Markets Department at SPsales@harpercollins.com.

A hardcover edition of this book was published in 2017 by Dey Street Books, an imprint of William Morrow Publishers.

FIRST DEY STREET BOOKS PAPERBACK EDITION PUBLISHED 2018.

Designed by Suet Yee Chong

Library of Congress Cataloging-in-Publication Data has been applied for.

ISBN 978-0-06-268493-6

18 19 20 21 22 DIX/LSC 10 9 8 7 6 5 4

For the love of God

"I play to people's fantasies. . . . People want to believe that something is the biggest and the greatest and the most spectacular. I call it truthful hyperbole. It's an innocent form of exaggeration—and a very effective form of promotion."

—Donald J. Trump, *The Art of the Deal*

CONTENTS

AUTHOR'S NOTE

This is a true story. It is also my story, which makes it a work of memory. To re-create what happened, I recovered photos, videos, text messages, e-mails, and social media—my own and other people's. I also relied on thousands of pages of reporting—a stack of papers more than a foot high—compiled contemporaneously by my NBC colleagues and me about the Trump campaign. Given that this is a story that unfolded during more than five hundred days, while I was in hotels, airports, TV studios, and more Trump rallies than I can count, I also hired a professional fact-checker. Where I re-create moments, I depended on videos, photos, and, where possible, conversations with the people involved. I cross-referenced my own memory with the memories of those who experienced the same moments alongside me. Most important, I took notes. A lot of notes.

Thank you to Donald Trump's friends, business associates, confidants, campaign advisers, staffers, and Republican

Party sources whom I agreed not to name. Thank you to the Trump supporters and protesters who were unfailingly polite to a reporter just looking to understand. Thank you to the Trump supporters and protesters who were anything but polite: you also helped me understand. Thank you to the Trump press corps. Thank you to everyone at NBC, from the interns to the executives. Without you, my reporting, and ultimately this book, would not be possible.

INTRODUCTION

If you've grabbed this book, there's a good chance you're still wondering how we as a country came to elect a former reality TV star to be our 45th President of the United States. Frankly, I'm still wondering myself. I bet you're also wondering: Can Trump do it again in 2020?

The book you're holding is the only unfiltered account of Trump's presidential campaign as it was experienced in real time. So, in retrospect, it's the best answer anyone can possibly offer (if I do say so myself). *Unbelievable* is my story of what it was like to be out on the trail in 2016 (for more than five hundred consecutive days). The questions then are still the same ones now: What is happening in America? Why is everyone so angry? Who are they angry at? Why are they cheering? Who is this guy they are cheering for? What do they see in him?

Of course, there are theories. (So many theories.) You can probably recite a few. There's the one about white people being afraid of becoming a minority in their "own" country. There's the one about workers being afraid that immigrants will take all the good jobs. There's the one about Democrats losing touch, and the one about Hillary Clinton being a "flawed" candidate. With even less-veiled sexism, there's always the one about how she wasn't "fit" for office. And I haven't even mentioned Russia.

These aren't the only theories and they won't be the last. The election of Donald Trump will be pored over by future generations, and we will never know for sure how Trump won the Republican nomination, and why he won the country. That's just life. It's a mystery looking forward and a mess looking back.

Still, I wrote a book. For some reason, people ask me to talk about it—a lot. Besides the constant question of how did we get here, there's the constant concern over how it's going. What is President Donald Trump doing? Is there a plan? Is this normal? Do his voters still like him?

People want to know whether I am surprised by the turn the presidency has taken. I'm not. Everything was foreshadowed on the campaign trail, including the Russia investigation. Remember Trump's flattery of Vladimir Putin? Or the news that Trump's campaign chairman had worked with Russian-backed candidates in Ukraine? Or the moment Trump invited Russia to find Hillary Clinton's e-mails? When I reread these scenes (in this book), the Russia investigation seemed not only unsurprising but also preordained.

Other behavior was easier to predict at the time. Specifically, the lying. Trump doesn't always err on the side of truth. That's not opinion. It is fact. The *Washington Post* has an ongoing tally of the lies, falsehoods, and misleading claims the president has made since taking office. As I write this, in June 1018, the number stands at 3,251. That's an average of 6.5 incorrect claims per day.

The president or those speaking on his behalf have lied or misled the public about small stuff, like the size of his inauguration crowd; weird stuff, like his hush money payment to a porn star; and really important stuff, like why the campaign and one of Donald Trump's sons met with a Russian offering "dirt" on Hillary Clinton.

Donald Trump built his political persona on the lie that President Obama was not born in the United States. He campaigned by stoking division using misleading crime statistics about minorities. He claimed that Mexico was sending rapists across the border and that Muslims were building bombs in their living rooms.

It shouldn't be shocking that he still lies or misleads. Just look at where it has gotten him. But what should be shocking, or at least confusing, is how Trump gets away with it. At the 500-day mark he enjoyed a greater approval rating among his own party than any other president after World War II—except George W. Bush after the attacks on 9/11. I'm talking about 87 percent of Republicans saying that he is doing a good job, according to Gallup.

Donald Trump is not just getting away with lying—he is thriving on it. When confronted with contradictions, he and his spokespeople bring up criticisms of their own, attacking the media and individual journalists by name, exactly as Trump did during the campaign. Why? It works for them. The harder he "counterpunches" the fact-checkers, the louder his supporters cheer.

A bit more surprising is the behavior of the Republican Party. Instead of checking President Trump's power, Republican members of Congress have made excuses for him ("He just does things differently.") or openly parroted his falsehoods (think Spygate!). Yes, there have been a few GOP lawmakers who have called BS (hello, Senator Jeff Flake!), but the vast majority of these profiles in courage are in reality profiles in retreat. They are retiring. No Republican primaries for them to face. No consequences for not back-slapping their man in the Oval Office.

The midterm elections may change the calculus. Then again they may not. Special Counsel Robert Mueller may find collusion. Then again, he may not. The lies may catch up to Donald Trump, or they may not.

The one thing that I'm sure won't happen in 2020—or at least I hope won't happen—is that the media won't underestimate Trump. That ended with 2016. While his political ascension was unbelievable, his ability to survive should be very believable. No matter the drama. Just as he repeatedly changed leadership on his campaign team, staff turnover at the White House is unprecedented. He's already fired two cabinet secretaries and, if you pay attention to his Twitter finger, it looks like he's itching to fire at least one more.

He's still weathering serious charges of racism, from arguing that there were some "very fine people" marching alongside white supremacists in Virginia to calling NFL players who kneel for the national anthem "sons of bitches" who "maybe shouldn't be in the country."

And then there are the women. In this #MeToo moment, where powerful men are being held accountable for a wide range of inappropriate behavior, it is notable that the man up top, the most powerful man in the world, has escaped consequence. Donald Trump has been accused by more than twelve women of sexual harassment and/or abuse. He has denied it. They are still accusing him.

So what happens next? Who will win in 2020? Pardon my French, but who the fuck knows? As they say in Paris, and as I'll say now, since I used to spend my weekends there before this all happened, *plus ça change*.

Katy Tur
New York, New York
June 2018

UNBELIEVABLE

TRUMP VICTORY PARTY

NEW YORK HILTON MIDTOWN
10:59 P.M., Election Day

I'm about to throw up.

I'm standing on the press riser at Donald Trump's New York City Election Night headquarters. Fox News is playing on two big-screen televisions, framing a stage covered with American flags and punctuated by two glass cases, each containing a MAKE AMERICA GREAT AGAIN hat. At the center, there's an empty podium gathering historical significance by the second.

"We also have a big call to make now," says Megyn Kelly, on the screen alongside Bret Baier.

As the clock strikes 11 P.M., the Fox camera pans across the studio to a jumbotron to reveal an oversized yellow check mark next to Donald Trump's grinning portrait in the state of Florida. Trump has just won it, along with all twenty-nine of its electoral votes. The ballroom crowd of staffers, supersupporters, and volunteers goes absolutely wild. The journalists in the room fall silent.

If the future is a blank sheet of paper, this news rips it in two.

My phone vibrates. And vibrates again. It doesn't stop.

"Holy shit, you called it!" flashes a text from a friend who had been insisting, like nearly all the polls on Planet Earth, that Hillary was a lock. I pick up my phone and check the *New York Times* election forecast. After predicting a Clinton victory for months, it has flipped. Trump has a 95 percent chance of winning the election, it says. Only two and a half hours ago, Hillary Clinton had an 85 percent chance.

Holy shit. I did call it.

In the seventeen months before now, I visited more than forty states, filing more than thirty-eight hundred live TV reports. I did all that as the Trump correspondent for NBC News and MSNBC, and I did it with one audience in mind: the American voter. My goal was to explain what Trump believed in and how he would govern if elected. The job came with all the usual hardships of the campaign trail plus a few new ones, such as death threats and a gazillion loops of Elton John's "Tiny Dancer," a staple of Trump's campaign rallies. I am proud of the work I've done but also quite ready for it to be over, thank you very much.

Ali Vitali weaves her way over to me on the crowded riser. She's been NBC's Trump embed since early on, a job that means not only attending virtually every campaign event, but also recording them for posterity. "Katy!" she says, with desperation in her voice. I am not prepared for the news she's about to deliver.

"Katy!" she says again. "He's going to keep doing rallies."

At first I don't understand her. He's going to be president—why would he keep doing campaign rallies?

"Trump," Ali says. "He's already planning victory rallies."

My head is a helium balloon.

Breathe.

The panic mounts.

More rallies?

I am nearly falling over.

More taunting crowds, more around-the-clock live shots, more airports, more earsplitting Pavarotti . . . I can't. I just can't.

The room goes wavy. My stomach churns. Lights flash in my eyes.

I'm never going on vacation. I'm never seeing my friends. I'm never getting my bed back. My brutal, crazy, exasperating year with Trump is going to end—by not ending at all. Trump will be president. The most powerful person in the world. And I will be locked in a press pen for the rest of my life. Does anyone really believe he'll respect term limits? I have a vision of myself at sixty, Trump at a hundred, in some midwestern convention hall. The children of his 2016 supporters are spitting on me, and he is calling my name: "She's back there, Little Katy! She's back there."

Anthony Terrell, my producer, taps me on the shoulder.

"They want you," he says.

I put in my earpiece and hear Brian Williams and Rachel Maddow digesting the news. In seconds, I'll be live in millions of homes. I can feel the bile in the back of my throat, but before I can swallow, I hear Brian building to a toss.

"Katy Tur is just up the block from us after a 510-day Trump campaign," he says. "What are you learning from there?"

Well, I've learned that Trump insists that he has "the world's greatest memory," but his vision of the future got him this far. I've learned that Trump has his own version of reality, which is a polite way of saying he can't always be trusted. He also brings his own sense of political decorum. I've heard him insult a war hero, brag about grabbing women by the pussy, denigrate the judicial system, demonize immigrants, fight with the pope, doubt the democratic process, advocate torture and war crimes, tout the size of his junk *in a presidential debate*, trash the media, and indirectly endanger my life.

I've learned that none of this matters to an Electoral College majority of American voters. They've decided that this menacing, indecent, post-truth landscape is where they want to live for the next four years. Look, I get it. You can't tell a joke without worrying you'll lose your job. Your twenty-something can't find work. Your town is boarded up. Patriotism gets called racism. Your food is full of chemicals. Your body is full of pills. You call tech support and reach someone in India. Bills are spiking but your paycheck is not. And you can't send your kid to school with peanut butter. On top of it all, no one seems to care. You feel like you're screaming at the top of your lungs in a room full of people wearing earplugs.

I get it.

What I don't get are the little old ladies in powder-pink MAKE AMERICA GREAT AGAIN hats calling me a liar. I don't get the men in HILLARY SUCKS—BUT NOT LIKE MONICA T-shirts. I don't get why protesting a broken political system also means you need to protest the very notion of objective truth. Because of Trump's war on the media, networks have required a traveling security detail except for Fox News (which hasn't been as demonized) and CBS (whose main correspondent is a guy who looks like he might be named Major—and is). A couple of weeks ago an advance staffer at a rally told me not to worry. "Save for Trump," he said, "you're the most watched person in the room. The Secret Service always has eyes on you."

I worry.

I also know enough not to mention it.

"The Trump campaign is feeling really good," I tell Brian, detailing what my sources are describing as the crazy, jubilant behavior inside Trump Tower at the moment. Trump himself has supposedly left. "He is upstairs spending some time with his family as the prospect of him becoming"—smallest of pauses—"president of the United States is suddenly a little more real than it was even earlier today."

I make it through the hit and the nausea passes. I have work to do, and nobody cares how tired I am. But that wave of *whoa* lingers. *It is unbelievable*, I think. All of it. Utterly. Inescapably. Completely. Unbelievable.

I'm writing these words on the eve of Trump's inauguration, seventy-seven days and at least seventy-seven thousand think pieces after Election Night. I've read countless articles about the 2016 election. Some have been insightful. Some have been absurd. As the very first national TV news reporter to cover the Trump juggernaut, I was there from the beginning—covering it every day for nearly two years, until the shocking end—and I've reached just one conclusion. Actually, two conclusions. First, no one can make perfect sense of it. Second, I'm smart enough not to try. The Trump campaign was the most unlikely, exciting, ugly, trying, and all-around bizarre campaign in American history. It roiled America and with it, my little life. I won't pretend to explain it. I will tell you what I saw.

1

"Katy Hasn't Even Looked Up Once at Me."

MAY 23, 2015
535 Days Until Election Day

Paris.

I'm up with the sun, in a studio apartment that's tiny even by New York standards. I think it is charming. The bed is in a loft, connected to the living area by a black iron spiral staircase. I climb down and tiptoe to the kitchen. In America, I'd make a giant cup of coffee, but I'm in Paris, where "filter coffee," as the Europeans call it, would be a sin and a spell breaker. I pop a little espresso pod into a sleek French machine.

I sip my espresso and stare down into the courtyard through a giant wall of windows, each panel of glass the size of a dining room table. The neighbors are chatting over breakfast. This isn't the Paris of tourists. I see gray tile roofs and smokestacks, not the Eiffel Tower. This is the real Paris, and I am an American cliché.

An NBC News correspondent dispatched overseas, I'm based in London but in love with a handsome Frenchman who is still sleeping up in that loft. The curtains are open, but Benoît has an extra set of blinds behind his eyelids.

You're lazy and French, I often say, hoping to get a flirty rise out of him. It always works. You're American and you are too much in a rush, he'll volley back in broken syntax. I'm smiling now at the mere thought of his snoozing face, the way he looks in the morning, rumpled and groggy, and the only person who gets to see it is me.

We met six months ago on Tinder. Yes, Tinder. I logged on while on a weekend trip with a girlfriend. He had a lapel mic in one of his profile pictures.

He's on TV. He can't be an ax murderer.

Ever since, my free weekends have all been Paris. The Eurostar from London is quicker than the Acela from New York to D.C.

I turn on French TV and try to laugh along with *Le Petit Journal*, France's version of *The Daily Show*. I understand precisely 3 percent of the dialogue but crack up anyway. The squawking television does its work. Benoît is up, looking down at me in disbelief, as if I've stirred him for a 3 A.M. fire drill instead of a 9 A.M. breakfast.

"I'm hungry," I say.

"Allons au marché!" he says.

"In English, please!" I say.

"*Mais*, Katy, you need to learn," he says.

"Pff," I say. "I know 'pff.' That's enough."

The rest of the day is one watercolor painting after another. The farmers' market. Fish on ice. Bright green lettuce. Small red strawberries. Eggs for breakfast. A scooter ride to Parc Montsouris. We collapse on a grassy patch. No blankets. No plans. An entire country with a ban on e-mail after work.

On the way home, we stop at a café for cheese and bread and a bottle of rosé. Benoît's friends Anaïs and François join us for another bottle—this time along the banks of the Canal Saint-Martin. Whatever you think of when you think of French beauty, Anaïs is it. François is in love with her, and I am in love with all of this.

K aty hasn't even looked up once at me."

The words boom through a microphone.

Huh?

I'm in New Hampshire just over a month later and it's raining.

Twenty feet across from me, on the other side of a backyard pool, Donald Trump is interrupting his own speech to scold me.

How does he even know my name?

"Lol. Trump keeps yelling at me," I text Benoît.

On July 11 Benoît and I are supposed to be in Sicily. The rooms have been paid for and so have the flights. Our first real vacation—two full weeks together. We'll swim in the Mediterranean, climb Mount Etna, and see opera in the ruins of an ancient Greek theater. And eat pasta. A lot of pasta. Part of me is already there.

The rest of me is right here in Bedford, New Hampshire. Katy Tur, Fearless Foreign Correspondent and Lady Who Drinks Wine at Lunch, is—for the moment, anyway—Katy Tur, U.S. Campaign Correspondent who, for no apparent reason, is getting called to attention by a reality TV show host turned presidential hopeful.

I came back to America because of a boy named Aaron, a severely ill teen who asked to shadow me for a day through the Make-A-Wish Foundation. Honored, I threw a few dresses into my carry-on and got on a plane. I didn't even bother to take my laundry out of the dryer. I left milk in the fridge. I'd be back in a week, I figured, wrongly.

On June 16, 2015, Donald Trump and his third wife, Melania, descended the Trump Tower escalator, waving to a cheering crowd padded with paid extras. Among the news media, the Trump announcement was seen as a sideshow. The headlines were savage:

DONALD TRUMP, PUSHING SOMEONE RICH, OFFERS HIMSELF.

FIVE FORMER PRESIDENTIAL CANDIDATES EVEN MORE RIDICULOUS THAN DONALD TRUMP.

DONALD TRUMP IS RUNNING FOR PRESIDENT AND IT'S GOING TO BE SO HILARIOUS.

Trump's speech added to the belief that he was not a serious contender. He said he'd be "the greatest jobs president that God ever created." He vowed to build a wall along the southern border and make Mexico pay for it. He delivered his opening lines with a frown and a scowl. His words did not seem destined for the history books.

When Mexico sends its people, they're not sending their best. They're not sending you. They're not sending you. They're sending people that have lots of problems, and they're bringing those problems with us. They're bringing drugs. They're bringing crime. They're rapists. And some, I assume, are good people.

Macy's dropped him as a business partner. Univision followed. The outrage built to such a point that NBC needed a reporter to cover it for a few days.

"Katy!" someone said. "She's just standing around."

I did a *Nightly News* segment and followed up on *Today*. At the end of my report, I told Matt Lauer and Savannah Guthrie that, despite all the anger, Trump was polling well in New Hampshire.

"He's number two," I said, "behind Jeb Bush." Then I reminded them, "It is very early."

"All right," said Lauer before moving on to the next story.

To understand how truly unexpected Trump was you have to understand something about presidential elections in general. The politicians devise strategies and court donors years in advance. At the same time, newspapers and networks carefully decide which reporter they'll match with which candidate. Trump wasn't part of anyone's plan. For that matter, neither was I.

Five days into my New York trip, while I was running an errand, I got a call from a friend at work.

"Hey, Katy. Heads up," the friend said. "Deborah Turness [my boss] is going to assign you to Trump full-time. [David, another boss] Verdi is going to call. If you don't want to do this, you better figure out what you're going to say to get out of it. Don't let on that I told you, but get ready."

Anxiety. Indecision. Italy.

My vacation with Benoît is in just over a week. On the other hand, as good as life can be in Europe, there's also a lot of professional boredom. It would be nice to get some TV time. And New York is unbeatable in the summer.

I hung up and paced the sidewalk. Then I called a friend from CBS.

"They want me to cover Trump full-time," I told him. My friend had covered Romney in 2012. "What do I do?"

He laughed. The whole thing was ridiculous: me following Trump, me on the trail, Trump running for president. Still, he urged me to do it.

"It will be fun," he said, "and if you hate it, at least it will be short."

A few minutes later, just as my source said, my phone blinked with a message from Verdi asking me to come see him back at 30 Rock. I didn't even make it to his office: he launched into his pitch in the hallway.

"How'd you like to spend the summer in New York?" he asked as we walked toward the elevators. Apparently Trump was not a *sit down in the office and talk about your future* kind of an assignment. More of a *let me tell you what you're doing as I walk to a more important meeting* gig. "We want you on Trump's campaign. It will be six weeks, tops. But hey, if he wins, you'll go to the White House."

He laughed. From everyone: so much laughing.

I said, "Sure." Or, rather, I heard myself say "Sure." In this business, your first answer is always yes. You can argue later.

"Oh, and you better get going," Verdi added as the elevator doors closed. "Trump has an event in New Hampshire tonight."

Political novice that I am, I decided to drive to New Hampshire. I figured I could make some calls on the road, and I like driving. Of course, any campaign veteran would get on a plane. There's a shuttle flight; you're there in a heartbeat.

My four-hour drive ended up being six hours, and I barely made the event. I pulled up as Donald Trump was taking the stage—really just a patio, with no podium. A kid in an ill-fitting suit introduced him to an audience of a couple of hundred overdressed people standing with umbrellas around a backyard pool.

Trump joked about his hair getting rained on. Then he launched into a speech focused on his unparalleled ability to run the country. "I get more standing ovations than anyone," he bragged. "So I have an expression that I use: the American dream is dead, but I'm going to make it bigger and better and stronger than ever before."

The crowd cheered. Then Trump paused.

That's when I hear my name.

"Here's my only problem. When television, I mean,

these people . . ." he says. "I mean, Katy hasn't even looked up once at me."

My producer elbows me. Trump is looking right at me. So is the crowd. So are the other reporters.

Me?

Trump and I have never met. Maybe he recognizes me from my days at NBC's local New York station. Or maybe he watched my recent pieces for *Nightly News* and *Today*. Or maybe there's something about me he likes. Regardless, he knows me, and he's now addressing me—first name only, as if we've been friends forever.

"I'm tweeting what you're saying!" I yell back.

And I am.

"At a campaign event in NH Trump repeats he loves Mexicans but that his statements about the border were true. #2016" . . . " 'It's hard to believe I'm second to (Jeb) Bush. This guy is not going to take us to the promise land.'—#trump #2016" . . . "Using a familiar campaign trope-Trump blaming the media for not getting the real story out. #2016"

Trump considers my excuse for a moment and nods approvingly.

"I hope so," he says. "I think you do a good job, by the way, Katy."

A man in the crowd yells out that Trump should buy NBC.

Trump doesn't disagree, and adds that he could "fix NBC." "I know what sells," he says.

The crowd seems to be refreshed by his whole performance. So I walk up to Trump's twenty-six-year-old communications director, Hope Hicks. She has bright green eyes and long, pretty brown hair. She looks like she walked out of a Lilly Pulitzer catalog. She greets me warmly and with impeccable manners.

"Can I get a pull-aside with him?" I say.

"Sure!" she says. "We'll do it right after he's done with what he's doing now." We trade numbers. I'm feeling pretty good. Then Trump closes his speech, walks straight to his waiting SUV, and speeds away.

Now what? Trump's gone. It's pouring rain. *Nightly* doesn't want me and neither does *Today*. My shoes are getting ruined. I consider calling Benoît. I still haven't told him about the assignment. I'm not so sure about the assignment myself.

I could say no. I have a perfectly valid excuse. NBC moved me to London nine months ago. I need to focus on that gig. Besides, it's not like they're asking me to follow Hillary Clinton or Jeb Bush. No one will wonder why I chose to go home instead.

Before I can dial Benoît, my phone rings and it's Hope.

"I'm sorry we left," she says. "Mr. Trump would love to speak with you at a later date."

A sit-down is considerably better than a pull-aside. I hang up feeling victorious. And that's when I realize that as much as I love my personal life overseas, I'm not completely satisfied professionally. My stories are fun but often frivolous—features that don't transcend "cute."

I FaceTime Benoît.

"Bonjour!" I say, trying to sound cheerful.

The connection is spotty. His picture is breaking up, pixelating, freezing.

"Hi, *babeee!*"

His voice is disconnected from his face.

I take a deep breath.

"*Tu me manques,*" I say in my shitty French. "*Mais . . .*" I don't know the words, so I blurt it out in English.

"I have to stay in America. I don't know for how long. I can't come to Sicily."

"*Mais,* Katy. You are a foreign correspondent," he says, pronouncing *correspondent* like fondant on a cake. "Zis ees not what we do *en* France." I can hear the anger in his voice. He's got a right to be angry. Our vacation is only a week away. But his anger annoys me. This is my job. I can't say no to my bosses. I can't tell them I don't want the opportunity. Who knows where it could lead? To put it another way, why be happy when you can be great?

I hang up the phone, not knowing when or if I'll see Benoît again.

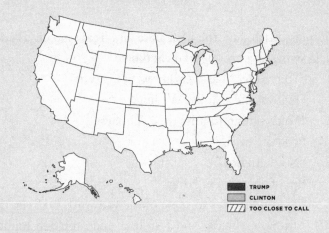

TRUMP
CLINTON
/// TOO CLOSE TO CALL

TRUMP-PENCE PRESS PLANE

GRAND RAPIDS, MI
11:59:50 P.M., Election Eve

"Ten . . . nine . . . eight . . . seven . . . six . . ."

All of us on the plane are counting down in unison. On the bulkhead, up where the first-class cabin would be on a normal plane, there's a large, rectangular clock. It doesn't show the time of day. It shows the time remaining until November 8, 2016. When we boarded, not so long ago but long enough, it was a big string of numbers.

"Five . . . four . . . three . . . two . . . one."

Do we cheer? Cry? Kiss the floor? It's Election Day, but our job isn't over yet. A minute later, we land in Grand

Rapids, Michigan, the fifth state and final stop on Trump's last swing as a presidential candidate.

"You ready?" I ask Anthony. "This is it."

He gives me a smile. We hug.

"You're going to miss this!" he says.

I roll my eyes.

"No. I am not."

But of course, I'm lying.

My mind flashes to some of the fun stuff. Like the time CNN's Sara Murray and I stole a midcampaign night in London, at the end of Trump's visit to Scotland. *"We're going to Shoreditch. We're eating at Dishoom. And we're getting two orders of lamb raan."* Or the time we danced to indie techno at Red Rocks in Colorado, laughing as Matt Hoye and Anthony tried to wiggle along to Jim Acosta's "dad dancing." Even the miserable stuff is kind of marvelous once it's over, like cheap birthday champagne, questionable airplane food, and even more questionable decisions. *Did you see them go up the elevator together? Does she know he's married? What happens in Vegas . . .*

I may even miss the deadlines, the calls—and calls and calls and calls—from *Nightly News* senior producer Eric Marrapodi: "They want to add Trump's tweet . . . We've got to cut twenty seconds . . . Sam wants a retrack."

Pop!

The memories are replaced with dread.

What do I do next?

I am afraid to slide back into general assignment reporting and paranoid about the loss of a mission, a purpose, a gin-clear reason to get up and go to work.

What if no one wants to put me on TV tomorrow? What if no one wants to talk to me? What happens when I'm not the expert? Will I still cover politics? Will I be happy to cover something else?

I can't think about any of this now. I push the questions out of my mind and grab my bag. We climb into buses and follow Trump's motorcade through the empty streets of Grand Rapids, toward the convention center. The bus is quieter than usual. We're exhausted. Whatever trip Jerry Garcia was on, it couldn't have been as long and strange as this.

"All right. Everyone out," says Stephanie Grisham, a Trump advance staffer who's been herding the reporters from rally to rally for months. She's our den mom. We love her even when she can't answer any of our questions (like where we're going tomorrow) and even when she has to explain, apologetically, that Trump has ditched us (he what!?).

The advance people are our only real friends on Trump's team. They're the only ones who spend time with us, and since they aren't exactly political hires, they're the only ones who don't mind getting close.

"Quickly!" Stephanie yells. "Quickly!"

We are more than an hour late. Or rather, Trump is more than an hour late. He usually lands before us and meets VIPs in a back room. We go in through the loading dock and emerge into a big concrete box already filled with Trump supporters. It's a Monday night—well, Tuesday morning, technically. Trump could've ended the campaign a few hours ago in

Manchester, New Hampshire, not here. It would have been a nice bit of symmetry: his big farewell on the site of his first big win. There were a lot more people there, too, and the advance staff arranged for a laser light show and smoke machines. It was like a rave where red MAGA hats replaced glow sticks.

But Trump needed to press on. We're here in Michigan because he needs a miracle. The polls look hopeless. (Sad!) So his team is throwing a Hail Mary in the upper Midwest. They think if any part of the so-called Democratic blue wall is going to crumble, Michigan is going to be the first crack.

Trump takes the stage as the folks on a riser to his right wave giant red letters that spell out his name. Ted Nugent played during the delay, a counternote to Lady Gaga and Jon Bon Jovi, the musical accompaniment at Hillary Clinton's last rally, going on right now in Raleigh, North Carolina.

"Today is our independence day," Trump says, which is almost a word-for-word line from a sci-fi movie. The crowd loves the show, but to me it feels just a little quieter than a normal rally, a little more still. A man in a Trump mask marches past the press pen. He's followed by a woman wearing a Trump lawn sign like a sweater. They're not leaving, but they are prioritizing an easy exit for later—a reminder that reality waits on the outside.

"Today, we're going to win the great state of Michigan," Trump says. "We are going to win back the White House."

I leave my phone in my pocket, for once, and do my best to savor the moment: my final rally. Day 500-something on the Donald J. Trump beat.

In twenty-four hours, I will be free.

2

"You'll Never Be President!"

JULY 8, 2015
489 Days Until Election Day

I wake up to sirens at 7:45 A.M.

Damn New York City.

Not sirens. My alarm. I reach for the snooze button just as reality hits: my Trump interview is this afternoon. I should get some coffee and read the papers before I go in. I scroll through my e-mails to see what's happened overnight. And there it is: a note from Hope. She's pushing up our interview, and not just a smidge. This is a four-hour adjustment, a swing from 2 P.M. to 10 A.M.

"Motherfucking fuck," I yell, jumping out of bed.

I really need to start waking up earlier. I also need to stop going to bed at 2 A.M. But right now I need to wash my hair, throw on a dress, and paint my face.

After a decade of television work and a lot of surprise

middle-of-the-night calls from the *Today* show—"Katy, sorry to wake you. We need you to front this weather/cop shot/lottery winner/missing girl/Hollywood gossip thing that broke overnight"—I can go from dead sleep to reasonably TV-ready within thirty minutes.

I hail a cab. From the back seat I e-mail NBC News's political director, Chuck Todd. Can he please meet me ASAP for that pep talk he promised?

"I'll come down right after the top of the Exchange," he says.

The Exchange is NBC's name for our morning editorial meeting. All the executives, show heads, senior producers, bureau chiefs, bookers, and any correspondent hoping to get on television that day must either be in that room or dialed in by phone.

I'll be missing it today. I check my watch: 8:30 A.M. I have an hour and a half to whittle down my list of twenty-one questions to ten.

Make that five.

"You don't know how long he's going to give you," Chuck tells me as we sip our coffee. "Don't count on getting more than a handful of questions in."

Chuck gives me a few other tips, all very helpful, like: "Try asking him if he owns a gun. Keep the questions short. Let him do the talking. See what he says."

Then he leaves me with a warning: "Remember, you're NBC. He's going to go after you."

It's now 9:10 A.M. My interview is a sit-down at the bar

in Trump Tower. It's just six blocks away from the office, so I walk over, Jimmy Choos be damned.

It's weird how calm I am. It's July in Manhattan, but I make the journey without sweating.

Trump Tower is a black, mirrored skyscraper that takes up half a block. The main entrance is campy, with an old-fashioned clock and white-gloved doormen in tails and gold-trimmed hats. More Disney than Dakota. I take a right on Fifty-Sixth Street and slip in a side entrance, which leads directly to the Trump Bar, where the "martinis" come in a wineglass with ice. Before I moved to London, I lived in New York for nine years and never once walked into Trump Tower.

Until now.

I know I've arrived when I see the thick black cables that will connect our interview to a microwave truck down the block.

The crew is two cameramen, a sound guy, the engineer in charge of that news truck, and two producers. The team has been setting up for hours. Everything is considered and reconsidered: the lighting, the background, the chairs. These guys are pros, and they have an arsenal of reassuring mantras: "If it doesn't work, we change it." "A network interview takes a village." "We make you look good."

The room takes on the feel of a reunion as I arrive. Crew members call out in surprise to see me back in the States.

"Katy!"

"What are you doing here?"

"What's going on?"

"We thought you moved?"

Hugs all around.

Now it's nine thirty.

And we wait.

Face powder. Notes.

Notes. Face powder.

I ask the crew questions: *Should I sit down? Do you want to mic me up? Do you need me for a lighting check?*

Hope Hicks appears, smiles, and gives us a two-minute warning. Trump is on his way.

Don't look nervous.

He lives and works upstairs, so I imagine he must have the timing of his golden elevator worked out to the second. At 10 A.M., right on time, he emerges into the lobby alongside a tall, white-haired man in a black suit. Trump is in a dark suit, the flaps of his unbuttoned jacket waving as he slowly covers the thirty feet of pink marble between the elevators and our location in the Trump Bar.

Certain people have a presence that's bigger than their physical size, an ability to ripple the air. They fill the room with significance, or at least a perfect imitation of it. Trump has that kind of presence.

And he's orange. There's no other way to describe him. He's the color of orange marmalade, perhaps a shade darker, like marmalade on toast. He adjusts his jacket and then adjusts it again in a losing effort to keep the flaps down and the tie in place. But he doesn't button it.

He also doesn't say hello, exactly, but sort of sings it.

He smiles and squints, and the sound seems to slip out the side of his face. His voice is lilting, almost cartoonish. We shake hands—and I go to take my seat.

Trump looks confused.

"Don't you want a picture?" he asks me, as if he doesn't know why I haven't suggested it yet. "Come here, Katy."

Okay, this is awkward. I don't want a photo. I know that our every move is beaming live into 30 Rockefeller Center, NBC News headquarters, and that my bosses, watching in real time, will cringe to see me smiling like a fangirl next to my interview subject.

I'm not sure it's a good idea to tell him no but at the same time . . . *why in the world would he think I want a photo? I'm not a fan. I'm a journalist. This is a network news interview.*

So I say yes. Maybe this is a mind game. Maybe Trump is trying to charm me, knock me off balance, confuse the point of this interview. Or maybe he just figures he's a big-shot celebrity and preinterview photos are routine.

The shutter clicks and captures my bewildered grin.

We sit down. The camera crew is lining up the shot and double-checking the lighting. They're almost ready when Trump calls for a time-out.

"Does that look good, Keith?" he says to his bodyguard, the man in the dark suit with the white hair. Keith is looking into the monitor checking Trump's shot.

"He's my stylist slash beauty consultant," he jokes.

I had heard a rumor that Trump was finicky about his appearance. In one version of the story, he hired a makeup artist away from a TV network because she was wearing gloves as she worked. Trump thought she must be hygienic. In another, that same makeup artist always told him to request "gold gels" from the camera crew. Thin pieces of gold-tinted plastic placed over the lights that she said would give him a rich glow.

Keith says his boss looks good, but Trump wants to check for himself. "Let me just see. Spin it," he says.

He scrutinizes his face on the screen and decides it's okay. The crew spins the monitor back around. Trump looks back at me.

"You know my whole life has been a win, you understand that," he says.

And so the interview begins.

Okay.

"My first question," I say. "Why are we here in New York? Why aren't we out on the campaign trail?"

"Oh, I've been to Iowa many times," he says. "I've been to New Hampshire many, many times. Love the people there. And we've had tremendous success. We've had tremendous crowds. Nobody gets as many standing ovations, and I spent a lot of time out. I was in South Carolina recently and we're all over. I'm going to, this weekend I'll be with Clint Eastwood in California, tremendous group of people . . ."

As a journalist, my job is to listen and probe, listen and probe.

As a human being, I'm struggling to identify every ingredient in his word salad. Is that a tomato or a radish?

After my top five questions are in, I ask five more, and then five more.

Twenty-nine minutes later, I've asked all my prepared questions, and I'm surprised he hasn't stopped me yet. Does he really want to keep talking? I can't tell. But I think I've got plenty for the producers to work with, so after he's finished answering my last question I say, "Thank you."

We shake hands and it's over.

My muscles start to relax.

That went okay.

Wait. Did it?

Suddenly, Trump is yelling at me.

"You better air that interview in full!" he says. "You're going to edit it. Deceptive editing. I know what you guys do. Deceptive editing!"

What is he talking about? Didn't we just shake hands? Did I do something wrong?

"It's not up to me how much of the interview gets used," I say, "but I know that we won't deceivingly edit you."

He isn't convinced. "If you don't," he says, "we have cameras in here; we'll release the full footage!"

Huh?

The threat is weird. How would he get the audio of the interview, for one thing, and where are the cameras? I look up and I don't see any, unless he means the security footage. More important, why would that scare me?

"You stumbled *three times*," he says.

He says it as if I killed a puppy.

"It doesn't matter if I stumble. I'm not running for president," I say.

What's with the hostility?

He looks me straight in the eye and lands what he must think is the harshest insult of all: "You'll never be president!"

Neither will you.

Thankfully, I bite my tongue before the words are out.

Hope Hicks interjects: "He's a presidential candidate. You can't speak like that to a presidential candidate. It isn't respectful."

Speak like what? What wasn't respectful? The man basically called all Mexicans rapists.

I'm baffled. I want to leave as soon as possible so we can get this interview on the air. Anderson Cooper's crew is at the door. Trump is sitting down with him next and I want my network, MSNBC, to beat his, CNN, to air.

I'm eyeing the exit, but Hope does not want me to leave. "Wait here."

What now?

She says Trump's campaign manager is coming downstairs to talk to me. This is the first I've heard of a campaign manager.

A small, skinny man with a buzz cut is walking over.

I offer my hand. He shakes it, quickly.

"Mr. Trump is very upset," he says in a high-pitched voice. I try to make sense of him.

How much do I need to care about this guy?

He looks like a sad salesman wearing his father's suit. He fidgets. He bites his nails. His eyes dart. They don't settle on me. He tells me I am unprofessional.

This is Trump's campaign manager? Did I hear Hope correctly? Why haven't I heard his name?

I smile and try to diffuse the tension. We should get a drink, I tell him. It would be great to hear what the campaign strategy is. (It's what you do, friends who've covered campaigns tell me. "Drink to befriend.") I'm still unclear on his job but I take his number and put it in my phone: "Corey Lewandowski—Campaign Manager??"

Hope and Corey leave.

My cameraman—a twenty-eight-year veteran—is stunned.

"I can't believe Trump yelled at you like that," he says. "He was so condescending."

One of my producers is laughing nervously. "Trump was really mad. Like, REALLY mad," she says.

Back at the office, my bosses are debating whether or not to air the interview unedited. "It's newsy from top to bottom," they say, "especially with how tense it got."

Everything is a fucking blur.

Tense? Was it really that tense? What did I miss? I can't remember any of it.

"Are you sure you were right about your stats?" one boss asks me.

"Yes, they're from Pew," I say.

"Are you okay with us just running it?" they say. "You did stumble."

I pause and consider it. One of the gifts of pretaped interviews is that you can usually cut that stuff out.

Eh, fuck it. What do I have to lose?

Chuck Todd is anchoring. The producers cleared thirty minutes of his hour-long show to play the interview in full. He introduces the sit-down, the tape rolls, and I take my place next to him. We're going to watch this together on live TV—me and the political director of NBC News.

Keep it together.

Trump appears on-screen, but I don't. In the rush to get it on the air, the editors didn't have time to mix in the camera that was trained on my face. So all the viewer and I hear is my disembodied voice.

"My first question. Why are we here in New York? Why aren't we out on the campaign trail?"

"Oh, I've been to Iowa many times. I've been to New Hampshire many, many times. Love the people there. And we've had tremendous success. We've had tremendous crowds. Nobody gets as many standing ovations . . ."

Always with the standing ovations.

His face is tight. He spits out his answers. He glares at me during the questions. He never smiles. Now I see what my producer saw. Trump is angry.

I hear myself move on to immigration, pressing him on the "rapists" comments. My research shows his link between crime and Mexican immigrants is flimsy at best. Mexico isn't "forcing" its "drug dealers" and "criminals" to cross the border.

"We have a lower incarceration rate for Mexican immigrants and illegal immigrants than we do for any U.S.-born citizen," I say.

"It's a wrong statistic," he spits back. "Go check your numbers! It's totally wrong."

He's trying to steamroll. Intimidate. Talk down.

"It's Pew Research," I say.

Now he's fuming.

Wow.

His rage didn't register in the moment. I thought it was all part of his shtick. The reality show star. But watching his face on-screen, it's clear Trump isn't playing.

He tries to tell me he's one of the most popular people in Arizona even though he hasn't been there in three years.

I ask him why we should believe him. After all, he led the birther movement. He had even claimed to have sent investigators to Hawaii to prove President Obama wasn't actually born there—which he was. We know because President Obama released his long-form birth certificate proving he was born in the U.S.

"If you believe that, that's fine. . . . Whether he did or not, who knows?" Trump says.

He rants about ISIS, and he calls himself a "big Second Amendment person."

"Our country is going to hell," he keeps saying. "We're being beaten at every front."

I ask him if he owns a gun. He says yes.

When I ask him if he uses his gun, he says, "It's none of your business."

I bring up his party. Republican pundits like Charles Krauthammer are calling him a "rodeo clown." He mocks

Krauthammer and appears to take a shot at his waist-down paralysis: "a totally overrated guy . . . a guy that can't buy a pair of pants."

He's going after a guy's disability!

He brags about his income and says he will release his financials. He calls the press dishonest. He says he'll win the Latino vote, "because I'm gonna create jobs." He rants about China stealing jobs. He praises Fox News for talking about immigration "bigly" (or "big-league," it's hard to tell which).

I ask if he is afraid of pissing off world leaders with some of his language.

He is furious. How dare I use that word—*piss*—with him? Rephrase the question, he demands. Okay. Are you worried you might "anger other countries?"

He cuts me off and calls me "naïve" when I flub a line. He condescends: "Try getting it out. I mean, I don't know if you're going to put this on television, but you don't even know what you're talking about. Try getting it out. Go ahead."

He's trying to get under my skin. I smile. Close my eyes, take a breath, and continue. He can't shake me. I'm not weak. If he wants a fight, so be it. He might like fighting. What he doesn't know is that I do, too.

Trump's celebrity doesn't impress me, and his combative style isn't new to me. I rather enjoy it, actually. It's familiar. My dad is opinionated. My friends like to argue. The people I love are all a little nuts. A major family meal without some-one storming out of the dining room is a failed gathering. No, not a lot fazes me, especially since, in my experience, the outraged party typically returns to the dinner table in time for dessert.

So when Trump blitzes, I sidestep. When he mocks, I smile. And when he gets angry, I assume he'll cool off. I get it: He's TV. This is his soundstage. I'm a part of his act.

I try to make small talk at the end of the interview, but Trump shrugs and then complains to my crew: "It's too bad that she only mentions the negatives. You don't want to mention all the positives."

"Wait!" I say. I tell him I have one more question.

He demands to know what the question is before he'll agree.

"I want to know why you think voters like you so much," I say, "and why you're getting such big crowds."

"That question I can handle," he says, to the crew's laughter. He puts his mic back on. "Because I tell the truth." Jeb and Hillary, he adds, "will never take us to the promised land."

I'm on TV again, this time live. I'm reacting with Chuck Todd. *Intimidating.* Chuck is the moderator of *Meet the Press*, the most notable and storied public affairs show in history, and Chuck is talking to me, the least notable and storied political journalist at NBC.

Either he senses my reverence or my face just looks funny, because the first thing Chuck asks is how I feel.

"I feel fine," I say, and I know what's coming next. The Trump campaign e-mailed a statement. Chuck reads it, shifting into his official *Meet the Press* voice. "We'd just like to reiterate our disappointment with the extremely negative positioning of the questions that were asked of Mr. Trump today."

Then it's back to me.

"Clearly he was not happy. Was it a public not happy, was it meant for us to see?" Chuck asks.

"It's significantly different watching it on the monitor than it is actually being there," I say. "That they were extremely unhappy with the interview when all was said and done is a bit of a surprise, to be honest."

I'm done.

Chuck moves on to the panel. The real politicos. *Hardball* host Chris Matthews. NBC White House correspondent Kristen Welker. Chief foreign affairs correspondent Andrea Mitchell. I did the interview, but I have no idea how to contextualize it for the viewer. For that Chuck needs seasoned Washington journalists. I am nothing of the sort.

"I think he is going for the fences," says Chris. "I think if you are sitting in an American Legion hall right now, or Knights of Columbus hall, or whatever, there will be a lot of cheering for that guy. They don't trust politicians."

"I kept having a flashback, Chuck," adds Kristen. "Reince Priebus. March of 2013. His one-hundred-page report after Mitt Romney was trounced by President Obama, saying the only way we are going to win in 2016 is if we reach out to people of color, to Latinos."

Chuck goes to Andrea.

My heart jumps a beat.

I've looked up to Andrea since my first day at the network. No one at NBC News works harder. *Today* show, *Nightly News. Today* show, *Nightly News. Today* show, *Nightly News.* She reports for both every single day. Not just that, but she anchors her own MSNBC show every single day. She knows everyone and she knows everything.

She proves it by fact-checking Trump's immigration claims on the fly.

"There is a net minus amount of immigration coming into the United States from Mexico," she says. "It's not just better enforcement: the Mexican economy has picked up, there are more jobs at home."

I'm impressed, but apparently so is she.

"A very strong interview," she says. "Katy Tur gave back as good as she got, really extraordinary television."

I'm dead. I can't believe it.

I step off set and check my phone. My Twitter account is on fire: *Who is this woman interviewing Donald Trump? He's going after her! He's attacking her! He's calling her naïve!*

I don't really know how to take it. Twitter is not my thing. What's the point of posting my work when all I get back are gross replies from men? A couple of years ago I tweeted a story about a brain-eating amoeba that killed a young boy in Louisiana. "Nice story on the brain-eating amoeba," some guy tweeted at me. "You looked really hot." That was the only reply I got.

Which is why today's attention feels so weird. No one is

commenting on my outfit. No one is rating me on my level of attractiveness. They are talking about the interview—the substance—and they are talking a lot.

My phone rings. Alex Moe, my old producer in New York, now NBC's eyes and ears on Capitol Hill, watched the interview with folks in Washington. The political press is praising me, she says.

"I told them you were terrible but they didn't believe me," she jokes.

I'm flattered and terrified.

I've been on TV for nine years, but no one knows who I am. I've covered tornadoes, terrorism, murders, fires, missing planes, epic floods, and devastating hurricanes. I've broken stories and led national newscasts. I've won awards. But people remember the news more than the name and reputation of the newscaster. This interview flipped that equation.

TRUMP-PENCE PRESS PLANE

GRAND RAPIDS, MI, TO NEW YORK CITY
1:51 A.M., Election Day

Have you ever wondered what would happen if you sat on a lunch tray and tried to "sled" down the aisle of a jet as it climbs?

We did.

CNN's Jeremy Diamond sits on two trays—one beneath his butt, the other beneath his feet. Our Boeing 737 is about to take off.

"Go!" someone yells. "Go!"

The flight attendant is already strapped into her jump seat. The booze cabinet, which has been locked all day, is open now and the alcohol is flowing.

The plane starts to accelerate.

"I want to wait," he says, holding himself in place. "I want to wait."

"Wait until we're getting erect!" someone yells.

"Fully erect," adds someone else.

He's smiling like a kid at the top of the big hill. Or a campaign reporter finally going home. And then it's happening, the force of the climb is trying to pin us to our chairs, and Jeremy puts his hands in the air and starts to slide. For a moment or two he picks up speed and a cheer builds. He's high-fiving people as he swooshes past, but then, after about eight rows, the ride is over. A dead stop. The plane is still ascending, but Jeremy's momentum is gone and now he's just sitting in the aisle like a dolt.

"CNN sucks!" someone yells. We all join in, a full chant. We're drunk—punch-drunk, sleep-drunk, work-drunk, and drunk-drunk, as we consume the last of the charter's liquor. Ali Vitali is crying. I met her a year ago by the Starbucks at Trump Tower. Her hair was short back then, pageboy short. She was sitting on a bench with her gear—a tripod, a camera, a laptop. She was stewing. She tried to put a good face on it, telling me how excited she was to work with me. But I knew she was unhappy about it. I wasn't a real political correspondent. Trump wasn't a real political candidate. And she knew damn well she was stuck on a junk assignment.

"Who would've thought?" she says, trying to laugh but also trying to wipe her eyes.

She's come a long way. NBC wouldn't have been able to break half as much news without her sources and tenacity.

She ditched her dumb New York City boyfriend (we've all been there) and found a nice boy on the trail—even if he can't aisle-sled as well as he'd hoped.

I pull her into me, sitting on the armrest.

"It's all going to be okay," I say mock solemnly while stroking her hair. She starts actually laughing and pushes me off.

A couple of rows back, Mark Halperin passes out in his seat. His mouth is open. His press credentials are still hanging around his neck. He joined us for the last few flights of the campaign while filming his documentary series *The Circus*. Now his producer and camerawoman are taking selfies behind him. CNN's Jim Acosta and I join in. A life-size cutout of Donald Trump appears—the next best thing to a candidate who actually shares a plane with his press corps.

I want to finish my beer, but it's getting late and *Morning Joe* wants me at 7 A.M., so I go back to my seat and try to get a few minutes of sleep before we get back to New York City. I put in my earphones and play a Phish song. I haven't really listened to the band in earnest since college. Now I find it is the only thing that can steady my nerves and slow my brain.

I drift off.

The wheels hit the tarmac of LGA at three thirty in the morning, shaking me back to life. In two and a half hours, America will start to vote.

3

"I Had to Grab Katy and Kiss Her."

NOVEMBER 11, 2015
363 Days Until Election Day

I'm exhausted and cold and, at 5:30 A.M., I've already been up for an hour.

The calendar may say midfall, but where I am the leaves died weeks ago. This is New Hampshire and I'm outside, dressed for television, not the weather. Anthony pulls up in a black SUV. I open the door.

Bless him.

He has the seat heaters on. My ass is happy and warm but my mind is a scribbling mess. The fourth Republican debate was last night, and I'm assessing Trump's performance. He was less confrontational—a surprise. In the last week Trump has been railing on Ben Carson, calling him a "hallucinating" "liar" with "insane theories." Why? Because

Carson is gaining traction in the polls. But for some reason, last night, Trump left Carson alone.

And that wasn't the only odd thing about Trump's performance. His foreign policy positions are still puzzling.

Take Russia.

In 2014, Russia annexed part of Ukraine, reclaiming a piece of the old USSR. Nobody in American politics condones the move and many Republicans count Vladimir Putin as a menace to world order, a threat to democracy.

Trump, on the other hand, presented Russia as an ally in the fight against ISIS and at best a secondary risk to American interests. As for Ukraine, and what some called Putin's "Nazi-style" aggression, well, Trump was blasé. Let someone else worry about it.

"We have a group of people, and a group of countries, including Germany—tremendous economic behemoth—why are we always doing the work?" he said.

It's not unusual for a new president or even a presidential hopeful to want to "reset" relations with Russia. George W. Bush said he looked Putin in the eye to get a "sense of his soul," but this is different. Putin's invasion of Ukraine isn't a staring contest. It's a land grab, the first of this scale since World War II. But when asked about it, instead of walking up to the plate and swinging at the softball (*crack!* more sanctions!), Trump put down his bat, walked down the third-base line, and kissed the opposing team's head coach.

"I got to know him very well," Trump said of Putin. "We were both on *60 Minutes*, we were stablemates, and we did very well that night." Pause. Smirk. Gesture to the moderator. "But you know that."

This answer made absolutely no sense. Stablemates? *60 Minutes* isn't *The View*. It's not live TV: no greenroom for guests to hobnob. And while Trump and Putin were both on the show, their interviews were pretaped at different times, thousands of miles apart.

Anthony was astonished. No serious Republican candidate would be so weak on Russia. "I hope someone asks him about that in the spin room," Anthony said during the debate. "If we were there, I would've told you to ask about NATO Article 5."

If we were there . . .

That's the other reason my mind is buzzing. We were not there. We were here, in New Hampshire. The only thing TV reporters hate more than round-the-clock live shots is handing over what should be *your* round-the-clock live shots to a colleague. Especially when your colleagues are trying to steal your beat. It's nothing personal. It's just the way news works. Everyone wants to be on the leading horse—never more so than when the leading horse is kicking, spitting, and bucking its way to the finish line.

"Are you sure New Hampshire is a better place to be today than Milwaukee?" I ask Anthony for the twentieth time since we got our assignment.

The marching orders came from our bosses in New York, though which bosses we could never really be sure. Every directive comes through what we call the assignment desk, a group of superorganized people who relay assignments, shrewdly laundered of their origin. I wanted to call someone to complain, but it would get me nowhere.

"Yes, this is the story now," Anthony says to reassure me. "I promise it's better." Ten minutes later, I ask him again.

I can't help it. I know the company never expected to have me on politics this year, and sometimes I feel like they don't know what to do with me. I've been on Trump from the beginning, but I don't have the story to myself any longer. Hallie Jackson is covering the rest of the GOP—which means she owns the news from six candidates to my one. Peter Alexander, one of our White House correspondents, is the *Today* show's go-to person for political spots.

To the outside world, Anthony keeps reminding me, I am *the* Trump reporter for NBC News and MSNBC. But while Anthony's optimism is boundless, it's doing nothing to ward off a creeping case of professional pessimism. MSNBC is still using me at the top of every hour, but *Nightly* is increasingly choosing Hallie, Peter, or Andrea Mitchell to wrap the Trump news of the day.

They know the scene better than I do. They've lived and worked in Washington. They know the players. The rules. The language: "doubled down," "baked in," "omnibus bill." *Omni-what?* "Here's the thirty-thousand-foot view," I hear Hallie say every morning, synthesizing the daily political landscape as if she had been doing it for years.

With each passing day, I have less of a lock on this beat.

"I've been there since the beginning" won't work forever. Political campaigns only get more complex. The further Trump goes, the less I'll be able to rely on being the outsider. I'll need to understand what it means to be an insider, too. And if all else fails, I need to break some news. Given enough time, anyone can become an expert on the past or public record. I need to be a reporter—a person who adds to it. I need to do it today.

We pull into the dark parking lot of Saint Anselm College. Anthony catches up on e-mail and overnight debate analysis as I slap on some makeup. Getting your TV face ready in the car is a skill that only years of practice can perfect. First, the lighting sucks. There's nothing flattering about a fluorescent overhead and magnet-triggered mirror LEDs. Second, the mirror is tiny. At most you get one-quarter of your face at a time. It's easy to come out looking like a knockoff Picasso: your image in four squares that don't quite align right. Too much blush on your right cheek. Not enough eyeliner on your left eye. The key is to build, lean back, build some more, lean back again. And repeat.

"Is it even?" I ask Anthony.

"You need more lipstick. Do you have anything brighter?"

"No."

He seems disappointed.

"Okay," he says. "There's a makeup artist inside. You can ask her what she has."

I hide my eye roll.

We walk into the "coffee shop," a student hangout transformed into a TV studio. Six director's chairs sit in a half circle facing three behemoth cameras on rolling tripods. This is the official reason we are in Manchester. Trump, who prefers to sleep in his bed in New York, has forgone that comfort, flown overnight from Milwaukee, and will appear on *Morning Joe* in less than an hour.

Mika and Joe are ready for him, sitting dead center in front of those cameras as a cast of political commentators and reporters revolve through the chairs around them. For a segment or two, I watch a merry-go-round of talking heads, set against a backdrop of young political nerds in hoodies, sipping coffee and waiting for Trump.

My hit is ten minutes away, so I check in with the makeup artist. She evens out my powder and sends me off to the set. I don't ask for brighter lipstick. My lips are fine.

I take my chair during the commercial break and say my good mornings to Joe and Mika. If you watch TV news, you might get the impression that we all keep chittering on through the breaks. Sometimes we do. But these road shows are barely controlled chaos. If the break ended just seconds sooner, you'd see.

Before I can ask, and without my consent, a guy has his hands up the back of my shirt. He doesn't need an "okay." He barely needs a "hello." The crew guys all know me. We've been working together for years. And this is just part of the job. He's got to string the microphone and other cables up my shirt—so the viewer can hear me and so I can hear the producers in New York.

"Hi, Katy, it's New York audio. Can you hear me?" someone says into my earpiece.

"Yes."

"Great. Can you count to five for a mic check?"

"One, two, three, seven, twelve," I say dopily. It's the same joke everyone makes, but for some reason I still think it's funny. It's probably not.

"Fantastic. Have a good hit."

I sit up straight, put a bright expression on my face, and wait for the introduction. We're "bumping in" with a sound bite from last night's debate—a line from Carly Fiorina, mocking Trump's suggestion that he met Putin in a green-room (even though *Morning Joe* found a clip where she had bragged about meeting Putin in the same way).

Great, I got this. I can say why Trump's Putin answer didn't make sense and then bridge into his foreign policy position, which seems to suggest he's a noninterventionist.

"Joining us now, Katy Tur, who has been covering the Trump campaign," Mika says, looking toward me. "Katy, how did you think the Trump campaign played last night?"

I don't jump right into the Russia stuff. Maybe I second-guess my knowledge. Or maybe I flinch, because I know Trump will be here any minute. I take up a more benign subject—Trump's shift in tone from previous debates.

Mika agrees with me.

"I do see a change," she says.

Now I'm running with it.

"What's his strategy?" Joe jumps in. "He's been pound-ing Ben Carson all week and then basically calls him a sociopath and then he, in the debate, seems to lay down the gloves."

I put my arm over the chair.

Casual Katy.

"A debate is a larger audience," I say. "He wants to come off a little bit more of the nice guy onstage. The guy that can bridge the gaps between the candidates."

We go a few more rounds on Trump's behavior and then I'm done. I don't get to Russia, or foreign policy, or Article 5, or his odd affection for Putin. The hit was fine, but fine is forgettable.

I get unhooked and go back to the greenroom for a cup of coffee. The makeup lady is cleaning her brushes so she's ready to powder the next riders on the *Morning Joe* merry-go-round. There is a fruit tray—melon, of course, no berries—and yogurt cups on ice, the morning TV version of bottle service.

Trump isn't due to get here for another twenty minutes, an eternity in TV time. So I squirt a cup of coffee out of one of those industrial canteens with the plunger tops. I stare into space and start to tick through my bona fides, still searching for some self-assurance. The last three months are a blur. I have forgotten what sleep feels like. In 120-odd days, I've been to more than forty-one different cities in at least nineteen states. Oh, and one other continent: I followed Trump to his Turnberry golf resort in Scotland, which allowed me to take a split-second detour into London.

I was there just long enough to have dinner, repack my bag, and have an ugly argument on a street corner with Benoît ("I'm too old to be having fights on street corners!" I yelled). We broke up. I've been too busy to mourn. It will make me too sad and make this job too difficult. So I'm not going to start now.

It's hard to find your self-assurance when you can't even find yourself on a map without assistance. It's also a law of TV: as your profile rises, your confidence tanks—you could plot it as one downward slope on a line graph.

And where are my clothes? *London*, I think. I can't remember what I packed, what I left, what I'll need. Everything is bought on the fly, ad hoc, a motley of styles and brands from whatever town I happened to catch a few extra minutes in.

I got the winter coat I'm wearing today in Atlanta during a rare midday break. It's big, army green, warm, and way too expensive. Not only is Anthony great at keeping the car warm and providing a makeup check, he's also a shopping enabler.

Maybe I am just feeling burnt out. Time off is getting harder to come by. There is no such thing as a weekend. I turned off my personal cell phone a long time ago. The guilt was just too much. I can't face three hundred unread texts from people I actually like, asking me about my life, telling me about theirs, probably wondering why I haven't responded.

One of the texts I actually read was from a friend who said she was pregnant with twins. Another friend said she was about to give birth. I want to see them. But Trump won't stop making news.

When I was chasing storms for the Weather Channel, I tagged along with scientists who would speed *toward* a spinning tornado. Even a quarter mile away, in the still silence at the foot of a twister, meteorologists will tell you that tornadoes and hurricanes don't scare them. Storms, for all their fabled unpredictability, still follow the laws of physics. The only thing that scares a meteorologist is lightning. There is no way to tell when or where it will strike. It is utterly unpredictable. Just. Like. Trump.

His controversies hit at all hours, at all times, and rubber soles on your shoes won't protect you.

TRUMP: John McCain isn't really a war hero.
TRUMP: Megyn Kelly is bleeding out of her "wherever."
TRUMP: Lindsey Graham's cell phone is 202-228-0292.

My "six-week assignment" is now four months of ever-deepening weirdness between Trump and me. After the "you'll never be president" episode, he's jogged between berating me and buttering me up.

He told a rally in Norcross, Georgia, last month, that I was "better" than most reporters. Seven days earlier in Franklin, Tennessee, he tried to introduce me to a crowd of hundreds of his cheering supporters, a hand on my shoulder like I was his wife. I spun from his grasp and walked away.

Then he was angry again. On my birthday, a couple of weeks ago, he singled me out in front of a rally in Atkinson, New Hampshire, saying that NBC reporter he won't name is terrible. Katy whatever her name, she's awful. Katy Tur is the worst.

He exaggerated the size of his crowds. I corrected him. He said he was self-financing his campaign. In fact, he was spending more of his donors' money than his own. He got booed at the Values Voter Summit in D.C. I asked him to respond.

"I didn't get boos," Trump told me as my camerawoman and I trotted backward trying to keep him in frame. "I got cheers."

Then he pointed at me: "I got boos for YOU!"

If Trump keeps dominating the polls, I'll have another year of this whiplash. Everyone back in New York and Washington seems to think Trump's days are numbered.

A big part of the country is already against him, they say, and now it's fall. The silly season is over. People are starting to get serious. They're paying attention. Football is back. *Homeland* is on. For heaven's sake, the Iowa caucuses are three months away! There's no way he'll make it.

But I don't see his demise so clearly.

Even after correcting for Trump's generous math, he's drawing unheard-of crowds a year before the election: twenty thousand in Mobile (he claimed forty thousand). Four thousand in Phoenix (he claimed fifteen thousand, then twenty thousand). Ten thousand in Springfield, Illinois. He said he beat out Elton John's crowd at the same venue—without a piano. True!

Trump supporters wait in lines that wrap around blocks. They say *The Art of the Deal* changed their lives. They ask Trump to sign dollar bills. Then again, almost all of them are white, which the leaders of the Republican National Committee have said isn't a large enough demographic to carry a national election.

Maybe the doubt is justified and Trump is doomed.

The thought cascades through my brain. Imagine the ability to wake up later than 6 A.M. To not have to sprint for a flight. To not have to read a dozen election articles before the sun is up. To take the Trump tweet notification off my phone. To have one live shot a day instead of eleven to fourteen. To go one complete spin around the sun without an anchor asking me every hour: *What have your sources been telling you?* As if I have a tiny Trump staffer who lives in my ear and is constantly feeding me new information.

'm halfway through my coffee when a producer says Trump is five minutes out. I know I have to say hello. But I want to stay here and hide. Trump is difficult and so is his staff. I've been dealing with them every day for four months, and it's rarely pleasant.

About a week after my July Trump interview, I went for a drink with a senior staffer. What did he think of Trump's chances of making it to the convention? I asked. "One in ten," he said. I told him about the unlikely way I got my assignment. He told me about his family back home. And with that I was ready to get on with my night. I looked at my watch. I had dinner with a friend.

I need to get out of here.

At the door of the restaurant, he had a question for me.

"Where can I go to meet thirty-something single women?"

"You have a wife and kids."

"So what?"

I laughed the way you laugh when your friend's grandparent makes a racist joke.

"I don't know. I'll see you later."

I tried to forget the exchange. Nothing I hadn't heard before. Also: *men*.

I don't know what the staffer thought after that. He was nice for a little while. He'd text back quickly, trying to answer my questions. But he wasn't entirely professional. He'd call at late hours, say disparaging things about women I worked with, comment on people's looks, claim well-respected female reporters were "fucking" this guy or that one. He'd tell me that he could prove it because he'd seen "text messages."

When the campaign sent water bottles and Trump towels to "sweaty Marco" Rubio, I texted the staffer to confirm it. His response: "You need some? I'm sure you get all sweaty sometimes too."

At a campaign stop in Waterloo, Iowa, he bragged to Anthony and me about all the women who would want to sleep with him when he became Trump's White House chief of staff. (So much for Trump's chances being "one in ten.")

As an embed during the 2012 campaign, Anthony had dealt with hundreds of staffers, and none had ever bragged about the sex they were going to have once their candidate won—especially when their candidate hadn't yet won anything.

"Is this guy serious?!" he asked when the staffer walked away.

Anthony is surprised, not naïve. We all know the phrase: wheels up, rings off. We all hear the stories. *Two reporters, married to other people, got busy under a blanket during an international trip on Air Force One.* We all get the warnings. *Whatever you do, don't sleep with a Secret Service agent.* We all know some people don't listen. *How do you think she knows where we're going before everyone else?*

I considered reporting details like this, but ultimately a campaign story isn't about the staff. Besides, I needed all the perspectives I could get on the candidate. And I thought my viewers did, too.

Again: *men.*

Trump arrives. Word spreads. I take a few steps toward the entrance and see Trump's private security detail, coming in just ahead of him. Despite my desire to avoid any interaction, I walk toward the front doors and spot Trump, who shifts his path ever so slightly so he's walking straight toward me—barreling, really.

Suddenly he is so close I can smell what he had for breakfast. And then, before I know what's happening, his hands are on my shoulders and his lips are on my cheek. My eyes widen. My body freezes. My heart stops.

Anthony looks at me startled, like, *What the hell was that?*

Trump lets go and saunters right onto the *Morning Joe* set, seemingly very proud of himself.

I'm mortified.

A week ago, *Good Morning America*'s Lara Spencer was

slaughtered for hugging Trump and *appearing* to sit on his lap when he visited the ABC morning show. The network said she was merely standing close to Trump, who was perched on a high stool. Regardless, the lap flap didn't help her credibility. And now here I am getting kissed by Trump.

Fuck. I hope the cameras didn't see that. My bosses are never going to take me seriously. I didn't have time to duck!

I look around for someone who will know whether the kiss made air. Trump is already on set talking about Ben Carson, who is threatening to overtake him in the polls.

"Is Ben Carson qualified to be president of the United States?" Joe asks, expecting Trump to do to Carson what he's been doing to "Low-Energy Jeb." But Trump isn't biting.

"Well, I don't want to say, Joe. That's not for me to say. I mean you know, look, it's, uh . . ." he responds.

"Would you let him run one of your companies?"

"Uh, I would let him operate on a friend of mine. Not necessarily me, but a friend of mine."

Trump smiles at his witticism. The crowd laughs, too. But Joe keeps pressing him about whether he'd let Carson run a company until Trump finally says, "Sure."

I find Jesse Rodriguez, *Morning Joe*'s ever-present senior producer. "Were the cameras pointed at the door when Trump walked in?" I ask, panicked.

"No, I don't think they were rolling until he got close to the set," Jesse says as Joe continues to press Trump a few feet away.

"I'm just curious why you're being nice this morning. Is it New Hampshire nice?" Joe asks.

"You know, I read where one of the folks said some really nice things about me this morning on someplace, and it didn't happen to be here." Trump stops himself midsentence, then looks at Joe as if he just had an epiphany.

"But actually, Katy Tur—what happened? She was so great. I just saw her back there. I gave her a big kiss. She was fantastic."

Jesus H. Christ.

Mika's eyes fly open. "Okay?" she says confused.

"I don't even know what to say about that," adds Joe.

A middle-aged man in the crowd is laughing open-mouthed. As the student at the next table picks up her jaw, she looks around the room and finds me. Her eyes are wide. She has the same uncomfortable smile as Mika. I turn and walk back toward the greenroom.

"Apparently he liked your analysis this morning," Anthony deadpans.

MY BED

HELL'S KITCHEN, NEW YORK CITY
7:28 A.M., Election Day

W ill you just shut up!"

  I'm alone, yelling at my morning alarm.

It doesn't listen.

  I'm fourteen floors above Times Square in my brown-toned corporate apartment: chocolate pillows, mocha blanket, coffee couch, beige art. I consider throwing my phone against the wall. It would make such a satisfying clatter, but it's got an indestructible case and the damn thing is going to keep beeping every two minutes for the next thirty minutes. So I give in, open it up, turn off the alarms, and check my e-mail. It's usually a jolt of news, a kick stronger than coffee or even a bucket of ice water. But Trump is quiet. No significant tweets. No news. I put my head back down and stare sideways into my closet.

What's clean?

It's Election Day and some of my colleagues are tweeting about their excitement—"eyes flying open" and whatnot. Not me. I am excited. Excited to be done. Excited to see how this ends. Excited to be a witness to history. Excited I made it! But I am not a morning person. I'm also half-sick or half-poisoned or both. The smoke from the machines at the Manchester rally made me cough all night. It was so bad that at 4:19 A.M. I had to give up on *Morning Joe*.

I e-mailed the show's executive producer, Alex Korson, and senior producer, Jesse Rodriguez. I'm sure it was the first time they had heard that smoke machine excuse, just as I'm sure they didn't believe it.

I'm anchoring at 2 P.M. today, then reporting for NBC and MSNBC for as late as it goes tonight, and doing both on three hours of sleep. I'll make it across the finish line, all right, and I will be smiling, but it will be more like a crawl than a sprint. I reach for the remote and turn on *Morning Joe*. Mika and Joe are introducing Kasie Hunt, who is near Hillary Clinton's house in Chappaqua, New York.

This should be interesting to hear.

Kasie is bundled up in a burgundy coat and white scarf. Her hair is smooth. Her makeup is on point. She's smiling. Although I know she is as internally ragged as I am, she looks like she just had a spa day. It's one of her secrets. What's not a secret? Her tenacity. In the early days, she popped in from time to time to cover Trump with me. We had a real bonding moment back in July 2015 chasing Trump down in Laredo, Texas, when we were both trying to get an interview with the candidate or an exclusive shot of him walking along the border. It was admittedly a little competitive and tense. I didn't want her to scoop me and she didn't want me to scoop her. We had it out right before Trump decided to hold a news conference so everyone could ask a question. If we were both going to cover him, we decided, we should strategize and work together. So we stood next to each other and tag-teamed questions. Kasie and I have been close ever since. She's a badass and a great friend.

NBC eventually moved her to cover Bernie Sanders, then Clinton. Kasie joined Kristen Welker and Andrea Mitchell on that beat. And when Trump seized the GOP nod, Hallie Jackson joined me. NBC calls us the Road Warriors. It's a silly name, but I like it. And I'm proud to be a part of it. We're the first female-led political team in network news history.

We didn't even realize we were making history until about halfway through the election when someone in the office looked up at the TV and noticed that everyone on-screen was a woman.

There was talk about marketing us as such. But Dafna Linzer—whose titles include but are not limited to managing editor of politics, meltdown soother, fierce ally, and (most important) friend—nixed that idea. "They're not the best women covering politics," she said. "They're the best reporters covering politics."

I couldn't agree more.

I turn up the volume on *Morning Joe* just in time to hear Mika say, "Kasie, how confident is Team Clinton heading into Election Day?"

Yes. This is the good stuff. I know how the Trump team feels right now—not great. But how does Clinton feel? Evidently, better—much better. Kasie says Team Clinton is preparing a victory speech and a concession speech.

"She was working on it on the plane yesterday with her speechwriter," Kasie says. "But that said, I think it would come as a real shock to this campaign if she ended up delivering that concession speech instead of the victory speech."

In that case, Team Clinton is feeling a lot better than it was just a few days ago. FBI director James Comey has reclosed the e-mail investigation that dogged her for months. She has a more than healthy lead in most polls. And everyone is saying she's a near lock as the forty-fifth commander in chief, the first woman president of the United States.

To memorialize this historic moment, her team posted a video of Clinton and her staffers doing the Mannequin Challenge on their final flight home. The idea is to make it look like time suddenly stopped. So everyone freezes in their place—like a mannequin—while a camera moves around the room. In this case it was the campaign plane. Two staffers are up front frozen mid–high five. Down the cabin others are drinking wine. Bill Clinton is near the back, smiling midconversation, just in front of Hillary, who is getting serenaded by Jon Bon Jovi. Everyone looks happy.

"This has really been, over the course of the last twenty-four hours, a celebration as much as anything else," Kasie says.

You can feel a *but* coming. "I will say, though, it's pretty clear that they are aware that there could be some issues on the other side."

She's talking about Trump. Not that he might win. But that he might not concede gracefully.

It's a perfectly reasonable concern.

4

"She's Back There, Little Katy."

DECEMBER 7, 2015
337 Days Until Election Day

Get her out," Trump says. "Security, strengthen yourself up. See, our country has this kind of security. That's the problem we have. Get her out."

TRUMP! TRUMP! TRUMP!

The crowd gets louder, and the candidate goes silent and glares at the protester.

"I don't want the person to be hurt, but I will tell you security is very weak. I can't believe these security people. One person. One person and we've wasted five minutes," he says. "Get 'em out."

The woman is removed, forcibly, and Trump continues.

"I wrote something today that I think is very, very salient. Very important. And probably not politically correct, but I don't care."

Here we go.

The entire press corps is listening now, waiting for Trump to address his call for a ban on Muslims entering the United States. Today is the seventy-fourth anniversary of Pearl Harbor, and Trump's team has decided to honor it with an incendiary announcement and a rally inside the USS *Yorktown*—a World War II aircraft carrier turned museum in Charleston Harbor.

The location is a surprise. Trump isn't speaking on the deck of the aircraft carrier. Instead he's in the belly of the warship: a big open space, used to display old navy bombers, fighter planes, and helicopters. The space is large but the ceiling is low, and the crowd has filled every inch of the floor like water, surrounding our press pen.

Yes, we are in a pen: a makeshift enclosure made of bicycle racks and jammed full of desks, reporters, and camera equipment. We're in the middle of the carrier, slammed against the right side wall. As usual, almost all of Trump's supporters are white and a lot of them are looking at us, not exactly kindly. The campaign and Secret Service force us to stay inside the pen while Trump is onstage. They even discourage bathroom breaks. None of them have a good explanation for why we're kept separate from the supporters. *Are we the threat or are they?*

"We're out of control," Trump says. "We have no idea who's coming into our country. We have no idea if they love us or if they hate us. We have no idea if they want to bomb us. We have no idea what's going on."

He narrows the focus to Muslims living in America. Do some agree with a global jihad? Do some want to be ruled by sharia law? You bet they do, Trump claims, citing some dubious polling data from an outfit the Southern Poverty Law Center classifies as a "hate group."

"By the way, I have friends that are Muslims."

Of course he does.

"They are great people. But they know we have a problem. They know we have a real problem. Because something is going on and we can't put up with it, folks."

Boos.

The sound is bouncing off the four metal walls of the ship, louder and louder.

"The mainstream media," Trump says. "These people back here, they're the worst. They are so dishonest."

Hoots and hollers.

And then I hear my name.

"She's back there, little Katy. She's back there."

Less than five hours before the rally, life feels like a dream of a different variety. I am at the best lunch place in Charleston, which puts it in the running for best lunch place in the South, which gives it a good shot at being the best lunch place in the world. You don't go to Charleston without finding time to get to Hominy Grill. You just don't do it.

"Who ordered the jalapeño hush puppies?" the waitress asks.

"Just put them in the middle of the table," I say.

"And the fried green tomatoes? And shrimp beignets?" she adds, raising an eyebrow.

Don't judge me, lady. Do you know how many gas stations I've eaten at lately?!

"Same thing. We're sharing," says Anthony.

Blessed Anthony.

He just arrived from a weekend with family, and I'm happy to see his face. I missed him. He is a political encyclopedia. He knows how lawmakers are going to react before they do and now he knows me, too. I swear he can tell when I'm about to blank on a fact or a name or a word. He'll literally call it out from behind the camera, just loud enough for me to hear it but not loud enough for the microphones to pick it up.

My mood is lifting. A lunch like this is a rare, hard-fought victory.

MSNBC has been asking for a live report from me every hour. The requests, once novel, now feel more like torture schedules, which is why we call it "the wheel of death." Each list of requests arrives at ten in the evening and then re-arrives at seven the next morning. It doesn't matter if you've negotiated with Beelzebub or one of his horned and hoofed minions the night before. The *morning* enforcers never get the message or don't care. You end up having the fight for time all over again.

This morning was typical.

"[Name redacted], we need a break," I said a little too forcefully.

No matter how much I tell myself to stay calm, no matter how hard I try to ask nicely, I lose it within seconds.

"For god's sake, I can't stand in front of a camera every hour until eternity! It's cold and it's wet and"—*Fuck, I've lost it*—"Jesus Christ, people need to eat!!"

"Katy, I'm just doing what I'm told! The shows have no one else. They're asking for you. Take it as a compliment."

"A compliment? I am going to die!"

I think Trump's melodrama is rubbing off on me.

"Listen, I'm sorry. I'm just hungry and tired. There's no food here. How about we skip the 3 P.M. live shot. If we can leave right after the 2 P.M. hit we can be back by 4 P.M."

"The four has been moved to four thirty."

"Even better. Thank you. Talk soon."

Ali is eating with us, too.

"And what about the shrimp and grits, the pulled pork sandwich, the low country chicken bog with stewed okra and collard greens, and the extra cheese grits, and the she-crab soup?" the waitress asks.

"Shrimp for her. Pulled pork for her. Chicken for me," Anthony says, pointing at me, then Ali, then himself. "Put the rest anywhere on the table."

Anthony's southern twang gets more pronounced when we cross the Mason-Dixon Line. UM-brella. ICED-coffee. IN-surance. He gives the waitress an ear-to-ear smile and a drawn-out "Thank you" and we dig in.

The food is stick-to-your-ribs, revive-your-weary-mind good. So is the company. I order the hummingbird cake for dessert. I don't even know what it is until it arrives and I love it. A banana pineapple spice cake covered in cream cheese frosting. I take a little too long to savor every bite and we have to hurry back to the battleship.

Two days ago, the mood was more of a nightmare. Trump called me out on his Twitter—four nastygrams in four minutes.

Dec 5, 2015 07:36:03 PM @KatyTurNBC, 3rd rate reporter & @SopanDeb @CBS lied. Finished in normal manner&signed autos for 20min. Dishonest! https://t.co/sCglbQjB3o

Dec 5, 2015 07:39:04 PM @Maddow, you copied incompetent @KatyTurNBC incorrect story. I'm sure you would like to apologize to me on show. Thank you for the courtesy.

Dec 5, 2015 07:39:42 PM "@Pimpburgh2015: @KatyTurNBC @realDonaldTrump just tweeted that you are a third rate reporter." That's only because I'm being nice!

Dec 5, 2015 07:40:17 PM @KatyTurNBC & @DebSopan should be fired for dishonest reporting. Thank you @GatewayPundit for reporting the truth. #Trump2016

Imagine someone calling you a liar. Now amplify the experience by a thousand if a presidential candidate calls you a liar. And tack on another factor of ten if that presidential candidate is named Donald J. Trump. Waves of insults and threats poured into my phone—the device buzzing like a shock collar.

He went on the attack as I was doing postrally live shots in an emptied-out barn in Davenport, Iowa. He was mad about something that had happened a day earlier, when he abruptly left the stage at a rally in Raleigh, North Carolina. (*Say "rally in Raleigh" ten times fast.*) For the first time, protesters got coordinated. There were only a couple dozen of them, but they took turns holding up signs that read STOP THE HATE and DUMP TRUMP and yelling things like "Black lives matter!" as loud as they could. Five or so minutes after the first group got thrown out, another one would start in. In total, they managed to interrupt Trump ten times in about an hour. By the fourth interruption, Trump had figured out their game.

"They've got a strategy," he told a crowd of about eight thousand at the J. S. Dorton Arena, home to political rallies, sporting events, concerts, and circuses.

After the tenth interruption, I tweeted that Trump had abruptly stopped his stump speech and walked offstage to shake hands. He denied this plain-as-day observation. Then he launched his Twitter attack. I should've seen this coming. Hope Hicks sent me an e-mail that morning.

On December 5, at 11:45 A.M., Hope Hicks wrote: "Katy, Mr. Trump thought your tweets from last night were disgraceful. Not nice! Best, Hope."

Check your phones. You'll get it in about six seconds if you haven't already," says Sopan Deb, the CBS News embed. Our beautiful lunch hour is about to end with a kick. Anthony went on to the USS *Yorktown*. Ali and I went back to the hotel to grab coats. Sopan is on his laptop in the hotel lobby. He's always on his laptop. Just as he's always wearing his oversized red headphones and what seems like the same maroon sweater and loose khakis.

Sopan gives us a warm hello and then hunches back over his computer. A moment later he gasps and contorts his face like a cartoon: popping eyes, hanging jaw.

"What?!" Ali and I say in unison.

Sopan doesn't respond immediately.

"Soap! WHAT IS IT?!"

His face snaps back to that bewildered smile, the one he gets when something completely absurd has happened.

We both look down, aggressively refreshing our in-boxes. And there it is: a campaign blast from Donald J. Trump. These can be anything from fund-raising e-mails to upcoming rally schedules. Sometimes they're meatier, but usually you have an idea about when those are coming.

This is different. The subject line reads: DONALD J TRUMP STATEMENT ON PREVENTING MUSLIM IMMIGRATION.

Is he serious?

I grab Ali's arm and we sprint to the car. I dial Anthony to tell him to get ready when my phone rings midcall.

"Katy, we're putting you through to a phoner line now."

The assignment desk: MSNBC wants me on the air now to explain the statement, *the unexplainable statement.*

"Katy? Are you there? We want to go to you now. Can you talk about this?"

"Okay," I say. "But it just hit, so I don't have reaction. I can just put it into context with his rhetoric so far—"

"Okay, just talk. Thanks, Katy."

I look at Ali, searching for an explanation.

"This is a proposed ban on all Muslims entering the U.S.?"

"Yup," she says.

A Muslim ban.

Breaking news chimes fill my phone. The anchor Kate Snow is saying my name, searching for the right way to set me up. "I guess the first question is how could that practically happen? How could this country shut down its borders to every single Muslim?"

"I'm not sure how that would happen," I say. "But this is just another example of the candidate becoming more and more hard-line, more and more extreme."

I do not have reaction yet—not from Trump supporters, not from the GOP, the Democrats, Muslim groups, not anyone. So I tap-dance. Kate does the same. The story is unusual but the muscles are not. We talk about recent polls and what Trump might think he has to gain with an announcement like this. I pull from all my experience on the trail, all my experience observing Trump. To me, he has a compulsive desire to be the best, smartest, and, in this case, toughest.

"He's really trying to position himself as the strongest when it comes to terror," I say.

Strong is one thing. This is another. Every time we think Trump can't go any further, he goes further. You can almost hear Jimmy Breslin say, "Beware always of the loud-mouth taking advantage of the situation and appealing to a crowd's meanest nature."

Except this time the subject isn't the Central Park Five, when Trump infamously took an $85,000 ad in four major New York newspapers, including the *New York Times*, to call for the restoration of the death penalty surrounding the case of five black and Latino teenagers—who were ultimately exonerated—for the rape and beating of a white woman in Central Park. This time Trump is talking about millions upon millions of people.

After the phoner, I start thinking about the deeper context. The closest thing I can come up with is the 1882 Chinese Exclusion Act. Congress suspended Chinese immigration for ten years, bending to demands from the American worker and growing fears of losing the country's "racial purity." But that applied to a country, not to an entire religion. There were immigration quotas after World War II but no outright bans of anyone for their specific faith—and Islam is a faith shared by almost one in four people worldwide.

I post a screen grab of the campaign statement to Twitter. The text is so far out there, the generalizations about Muslims so broad—"*it is obvious to anybody the hatred [Muslims have for America] is beyond comprehension*"—that some on Twitter don't believe it is real.

This is the end, the pundits say. This is surely the end.

Close to 4 P.M., Ali and I pull up to the dockyard in front of the USS *Yorktown*. Trump is scheduled to be there in about three hours, and supporters are starting to line up. We need to get their thoughts, but I also need reaction from the GOP. So I stay in the car and on the phone while Ali takes the camera crew to the line that's forming at the foot of the docks.

I use one phone to talk to MSNBC. I use another to text and call sources. Everyone in TV news has two phones. Add in your personal phone and some of us have three. You forget how weird it looks until someone tries to casually ask if you're a drug dealer.

In a matter of minutes, Ali speaks to dozens of Trump supporters, and we get what we are looking for, the beginnings of an answer to the question of questions: Has Trump gone too far? We know traditional Washington alongside "the media elite" will immediately condemn Trump and declare his candidacy sunk. The same people, saying the same thing they've been saying since day one. Trump is a clown. Trump is dangerous. Trump is doomed. What we don't know is what will happen when people actually vote. But by midafternoon I have what feels like the future right here in my inbox, courtesy of Ali Vitali.

No one she spoke to is disturbed by the Muslim ban.

"It's a wise decision," said one man waiting in line.

Another man, a soldier who had done tours in Iraq and Afghanistan, went further. To continue to allow Muslims to come in would be a "kick in the face to every veteran," he said. The only thing better than a ban would be mass deportations, he said.

"Ship them all back."

W alking along the gangplank to the warship, I call *Nightly News*. In the craziness, I hadn't had time to check in. But it doesn't matter. They don't want me in the broadcast that night. Apparently, Andrea Mitchell is going to handle the spot from Washington. She has the foreign policy chops, after all, and *I'm just the asshole at the venue who's been following this guy for six goddamn months.*

I don't blame them for wanting Andrea. I'd want her in this situation. No matter how good I may think I am, the bosses will go to experience first. When it comes to policy and its international effects, Andrea is the queen. She's dedicated decades of her life to it.

But a wave of frustration crashes through me. Washington's reaction isn't the whole story. I'm on the ground. I know Trump. I know his base. They may need Andrea, but they need this perspective, too.

I see Anthony, holding a spot for me in the security line.

I explain the *Nightly* situation.

This time he doesn't have a good consolation. There isn't one.

"Call Janelle," Anthony says, referring to Janelle Rodriguez, the SVP in charge of *Nightly*. "Tell her it's *your* story. It's your beat. You are here. And you have the reaction."

I call.

Janelle understands, but she makes no promises about getting me on TV.

As soon as Ali and the crew got the interviews fed back to New York, the team started breaking down the live-shot position. No small talk. No cigarette breaks. It was a controlled riot of activity, a rush to get all their gear locked up and for me to get inside for prime-time coverage. A statement from a politician—even a statement as unvarnished as the one this afternoon—is never as revealing as a speech with the same idea. Everyone is anxious to find out if Trump will soften the edges or sharpen his knife.

I am in a rush now because of security. Since Trump got a Secret Service detail after Thanksgiving, our days have become infinitely more frustrating. For one thing, access to him is now nearly impossible. Reporters used to be able to walk right up and shove a microphone in his face—which he rarely ignored. Now dozens of armed, black-suited men and women with radios in their ears and microphones up their sleeves surround Trump at all times.

Second, we can't just post up inside the venue from dawn to dusk anymore. There's a strict protocol. The crew has to show up five to eight hours early to "preset" their cameras, tripods, sound equipment, and cables inside the arena, auditorium, theater, farm expo, or, in this case, warship. The

Secret Service then kicks everyone out and spends three to five hours "sweeping" the gear with bomb-sniffing dogs and other fancy gizmos and gadgets.

By the time they open the press doors for reentry, anywhere from a couple dozen to a couple hundred journalists line up—because we have to get screened ourselves. Secret Service agents—often with help from TSA agents (no joke)—unpack our bags and "wand" us with handheld metal detectors. The Secret Service only has so many agents, so the TSA steps in to help at particularly large rallies. Is this part of the reason the lines at airports seem so much longer nowadays? Not enough staffers to work the security line? I don't know for sure, but c'mon.

The agents are just as frustrating as they are at the airport. Sometimes they'll confiscate my oranges (I could throw them at Trump) or my dry shampoo (I could light it on fire). They'll *try* to confiscate the hair spray that Fox News's Carl Cameron always has on him. But after decades of dealing with presidential campaigns, he's come up with a surefire way to get it through. "You taking my hair spray is like me taking your gun," he explains. This the agents can understand. They don't call him Campaign Carl for nothing.

Janelle calls back about *Nightly* at 6:10 P.M. Twenty minutes to airtime.

"I got you in," she says. "You're doing a cross talk after Andrea. Give us the context you've been adding on MSNBC.

How did he get to this point? Send us an e-mail with what you're going to say. Don't let me down."

By the time we hang up, it's 6:15 P.M. I've got to pee and I've got to write a sixty-second cross talk. Freewheeling on MSNBC is easy. The shows are fluid. *Nightly* is not. It's a twenty-two-minute broadcast. Each report is timed down to the second. If you don't stick to what they give you, you'll throw the whole show off. Which means you have to weigh every word you say.

I'll pee later. I have to get something down.

I've done thousands of live shots in my career, but whenever I hear Lester Holt say my name I get nervous. I can't help it. The broadcast is still a big deal: eight to ten million viewers a night, many of them in the middle of America. It's a broadcast that carries weight. But it is more than that. It's history. When people go back and research this election, they will pull up newspaper articles and network broadcasts. Our words are recorded and remembered. They live on long after we die.

I write out what I am going to say in a reporter's notebook. Anything longer than two pages is too long. This is three pages. I read it out loud to hear how it sounds, then I go to work on it, crossing out whole sentences, shortening others. When I think I'm close, I time myself on my iPhone. Then I cut it down further, knowing that my senior producer is bound to get in my ear thirty seconds before Lester tosses to me to say I only have forty-five seconds.

When the copy is final, I write it down cleanly. Then do it again. I did the same thing in my high school history classes. The physical act of writing helps me lock words into my brain. I practice a couple of times, looking into the camera, then try to clear my mind. If you hold on too tightly you'll screw it up. You have to trust yourself. Lester says my name and I go:

"This is just the latest in a long line of extreme comments as he's become more hard-line on the campaign trail. First it was a Muslim database, then surveilling Muslims, then closing all mosques, now he's saying a ban on Muslims coming into the country, including tourists, and even Muslim Americans living abroad.

"His supporters tell us they like this. They believe it is a wise decision. And they believe Donald Trump is going to keep them safe. Why? In our latest MSNBC poll we found that 60 percent of Republican voters say that one of their biggest concerns is being the victim of a terrorist attack."

Inside the aircraft carrier the mood is dark long before Trump takes the stage. The air around me feels somehow flammable, like the air around a gas station. I am afraid that Trump himself will strike the match.

No one has ever called for a ban on an entire religion before. Even after the horror of September 11, George W. Bush made a point of standing on the rubble of Ground Zero and soon after imploring people not to take this out on their Muslim neighbors. He made it clear that terrorism was a perversion of Islam, that the vast majority of Muslims were peaceful.

Trump is implying something quite different, and stoking fears: *Muslims are scary. They're different from you. You're right to be scared. There are terrorists in your neighborhood. Your Muslim neighbors are hiding them.*

We're standing a few miles away from where a white kid walked into a black church, Emanuel AME, and killed nine people. But that's not a threat that seems to compute in Trump's world. A group of innocent black people were murdered by a deranged young white man. That happened in a whole different political universe.

At 7 P.M. Trump takes the stage. The room shakes alive, roars as one.

"Wow, thank you. Thank you so much," Trump says with a smile. His mouth seems to have two positions. One is a perfect oval, where his words seem less pronounced than ejected. The other is a straight line that cuts his face in two. No teeth, lips stretched. Self-satisfied.

His supporters whoop and cheer. Trump pulls a stack of papers from his suit jacket. "We start by paying our great respect to Pearl Harbor." He looks down at his notes. "We don't want that stuff. We don't want World Trade Centers. We don't want that ever happening to us again. It's not gonna happen." He adjusts the mic, thanks the crowd, claims there are thousands waiting outside, and then launches, as he always does, into his poll numbers.

Some politicians have a gift for language. Trump is not one of those politicians. His sentences call to mind an aerial shot of a burning, derailed freight train. The syntax is mangled. The grammar is gone. "Donald Trump isn't a simpleton, he just talks like one," reads a *Politico* article from last

August. "If you were to market Donald Trump's vocabulary as a toy, it would resemble a small box of Lincoln Logs."

Every fourth word seems to be *very*, *great*, *beautiful*, or *tremendous*. He loves the word *winning*. In fact we're going to have so much of it, Trump says we'll get sick of it. His insults are even simpler. Our leaders are "dumb," "stupid," or "weak." Our deals are "terrible." His critics are "losers" and "haters." The press is "scum." Women he doesn't find attractive are "disgusting."

He is the polar opposite of President Obama. Where Obama's rhetoric soars, Trump's rhetoric slithers. While Obama eats arugula, Trump scarfs Burger King. Where Obama is controlled and calculating, Trump is petulant and loud.

But it works for him. Everything he says falls into one of two categories. If it's good it's "we." If it's bad it's "they." "We" are going to have so much winning. "They" are going to hate it. His supporters feel that he is fighting for them. They identify with him. They can relate. "He talks just like us," supporters say over and over again. He's the rich guy they would be if they were rich. And he knows it.

The crowd is laughing now. Trump is riffing on ISIS, his opponents, complaints about his "tone."

He does this routine all the time. But tonight it has new weight. I'm sitting on the riser, below my camera, taking notes as he talks. I can't see Trump's podium from my angle, but I don't mind. I've seen enough of his face. I can picture in perfect clarity every scowl, smirk, duck face, jaw drop, and gesticulation he is making. I know his speeches by heart. I know the whole act. I've been living it for six months. Besides, something tells me tonight is a night to lay low.

Suddenly, a protester interrupts the speech. The crowd fires off a fusillade of boos. Trump seems angry. On the monitors, I can see that he walks away from the mic and tells his personal security guard to get her out. Meanwhile, the protester is still yelling. Her voice ricochets off the ship's steel hull. Trump is visibly annoyed now. He walks back to the microphone and admonishes the local South Carolina police officers assigned to keep an eye on the rally.

Trump resumes talking and I keep my head down—until I hear my name.

"She's back there, little Katy. She's back there. What a lie it was. No, what a lie, Katy Tur, what a lie it was from NBC to have written that. It was a total lie and they did a story where they said I didn't know they had a group like this, where they actually criticized the media. And they said it was a total lie. And I loved it, I loved it, I loved it. And then other people pick it up, you know, it's NBC, so somebody picks it up. Third-rate reporter, remember that. Third rate. Third rate."

My heart stops. My lungs clench. Every camera in the press pen is broadcasting the rally live on TV or online. Millions are watching at home to find out how Donald Trump is going to justify his Muslim ban. They also get to find out what he thinks about yours truly.

It's clear what the crowd thinks: they love it. They turn

all at once, a large animal, angry and unchained. I force a laugh.

Shake it off. It's worse if they think he scares you. Just smile. Smile and laugh.

My face obeys. I throw in a wave for good measure. But inside I'm terrified. Men are standing on their chairs to get a look at me. They want to see me as they jeer. An older woman to my left is horrified. A friendly face in a crowd of thousands. I decide to tweet about it, hoping my nonchalance will project strength.

I'm not going to let this guy get into my head.

Once again, my phone is going nuts.

Journalists are tweeting in my defense. My bosses are texting to find out what happened. My mom is borderline hysterical. "Are you ok?!?!?"

I switch it to silent. The constant vibrations are distracting, and Trump is finally getting back to the ban.

"We put out a statement a little while ago and these people are going crazy," Trump says, pointing to the cameras. He literally waves the criticism off. The crowd chuckles.

"Donald J. Trump is calling for—and you gotta listen to this one, because it's pretty heavy stuff and it's common sense and we have to do it."

He is relishing the moment and so are his supporters. They are spellbound. He looks down at his notes and reads verbatim from his press release:

"Donald J. Trump is calling for a total and complete shutdown of Muslims entering the United States until our country's representatives can figure out what the hell is going on."

He adds the "hell" for effect. The room explodes.

Hardball wants me live. I take a deep breath, stand up, put in my earpiece, and hook back into MSNBC's live coverage.

"Well, let's go to Katy Tur. Katy, are you used to this kind of trash talk from him?" I hear Chris Matthews but I can't understand what he's saying. Trump is still bellowing behind me. Chris tries again. "I'm trying to couch this in the most politically correct way. Are you used to the trash talk that Donald Trump threw at you tonight?"

I still can't understand him. It doesn't help that Chris tends to talk a mile a minute. So I just go, deciding to ignore Trump's comments about me and just address the ban.

"Thank you so much, Katy Tur, in South Carolina," Matthews says.

I stay hooked in to MSNBC. Trump is done. Chris Hayes is taking over in a couple of minutes. He wants me on his show, right at the top. Supporters are taking their time to leave. They're still whipped up. I know someone is going to yell at me as soon as I start talking. So I do what I always do. I find the pinhole deep in the back of the lens and I tune everything else out.

A couple of minutes later, I'm done. The crowd that gathered behind my live shot is gone except for a few stragglers, yelling at me. They're five feet away—held back by those lousy bicycle racks. A Trump staffer shoos them away. MSNBC has cleared me and my bosses want Anthony and me to get out of there as quickly as we can. I don't quite understand why until we pack up and start to head out. A

Trump staffer stops me and says, "These guys are going to walk you out."

I look over and see two Secret Service agents. Thank goodness. They walk Anthony and me along the gangway back to our car. It's pitch black and I'm nervous. We're parked with the crowd.

Once we're moving, I take a look at my phone. My mom has called. And called. And called. I dial her back. "Are you okay? Where are you staying? Can someone stay with you? You need security!" She is crying. And it hits me.

I'm a target.

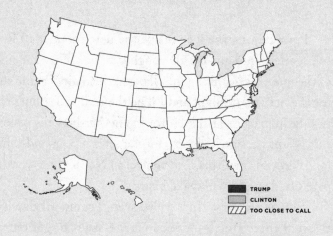

NBC NEWS HEADQUARTERS

MIDTOWN, NEW YORK CITY
11 A.M., Election Day

Trump steps from a black SUV. He waves and smiles at the gathered crowd held back by barriers.

Who knew he had so many fans in New York City?

I turn the sound up.

"Boo! Boo! Booooo!"

Oh.

Melania, Ivanka, and Jared follow Trump out of the car. More boos.

His wife, daughter, and son-in-law are going with him to vote. They look like the picture of success.

So long as the TV is on mute.

I'm watching the feed inside the newsroom at 30 Rock, four coffees into my day, thoroughly awake. The excitement of the moment is sinking in. I'll be on the main Trump story for the network tonight and I'll be the lead at Trump's election headquarters at the Hilton. Whether he wins or loses, this is a professional victory for me after a year-and-a-half-long fight to own this beat for NBC News. The work has paid off. Tonight will make history.

This is the first time we've seen Trump all day. He's tweeted three times—innocuous messages to get out and vote—and granted a wandering phone interview to *Fox & Friends*, in which he padded the crowd estimates at his last campaign rally by about seventeen thousand people and said a loss tonight would be "a tremendous waste of time, energy, and money." But this footage of him voting is our first glimpse today of the person I, despite the polls, believe will become the president-elect sometime this evening.

Inside the school, Trump buys a cookie from a couple of little boys and gives them each a high five. Then he hands the cookie to Melania, who's in dark glasses and a chic white dress, a camel coat draped over her shoulders like a cape. From the expression on her face, he might as well have just handed her a dripping hot dog.

"I'm going to eat that later," he says. Trump gives the boys a twenty and starts for the gym. He doesn't make it two steps before an old woman with a walker blocks him. She takes his hand and tries to pull him in close. He stays put, so she hangs on.

"Ohhh, I'm so glad to see you," she says, cooing over the candidate.

"Did you vote?" Trump asks.

I can't understand her answer, but Trump appears to like it. He grins and pats her on the hand. Before Trump arrived, two women took their tops off and yelled, "Grab your vote," a reference to what Trump has said his celebrity allows him to do when he sees beautiful women ("grab 'em by the pussy"). NYPD officers carefully removed the women and charged them with electioneering.

The only remaining hubbub is from photographers and cameramen, jostling to get a better shot of the candidate. Trump walks toward the press pool and pauses for half a second. He's wearing a black suit, white shirt, and long blue tie. His overcoat is on. He's been wearing it everywhere, even inside at some rallies. His shirt is boxy and bulky from the bulletproof vest he started wearing everywhere back during the primaries. He's bigger than he was when the campaign started, a little more pear shaped. Melania, on the other hand, looks as polished as ever. Her hair is hanging over her face. I wonder if it serves as her own armor. Maybe it gives her a little privacy in the face of the crowds she can no longer escape? She keeps her sunglasses on, even though she is inside. Maybe that helps, too?

If Trump was considering stopping to talk to the press, he thinks better of it. He and Melania keep walking. The cameras are clicking and snapping.

"Mr. Trump, what are you hearing about early returns from your team?" Jill Colvin of the AP shouts out. Trump doesn't break step; instead he turns his head back and says, "Very good."

Except it doesn't look very good for Trump. NBC's final battleground map shows Clinton likely getting 274 electoral votes to Trump's 170. Print reporters are taking bets on the tight Senate races, ignoring the top of the ticket. The outcome is too obvious.

The voting booth is more of a voting desk with cardboard privacy screens around it. Trump has to bend down to fill out his ballot. Melania is next to him. Her sunglasses are now off. She, too, is bent over her own desk, her own ballot. A moment later, Trump pops his head up. He looks at his wife. My phone buzzes with Twitter notifications. "Did Trump just look at Melania's ballot to make sure she was voting for him?" the Internet seems to ask all at once.

Outside, the boos are even louder as Trump walks back to his car, except now there are some cheers. A group of construction workers showed up. One calls out, "Lock her up, Donald! Lock her up!" Trump waves and smiles. Someone else yells, "Loser!" Trump pumps his fist for all of it. Attention is attention.

5

"Katy, It's Donald."

DECEMBER 21, 2015
323 Days Until Election Day

Franklin Roosevelt was the first president on television. He appeared live from the World's Fair in New York in 1939. This wasn't mass media: just a few flickering moments, on a few experimental television sets. But some of those sets weren't far from where I am now, inside Rockefeller Center, which hasn't changed much over the years. The buildings span three blocks and include plazas and rooftop gardens. The centerpiece is an iconic seventy-story limestone tower, capped by a rooftop observation deck.

The big difference is stature. Back in 1939, the tower was known as the RCA Building, and RCA—aka the Radio Corporation of America—had virtually every aspect of radio under its control, thanks to a massive then-new division called the National Broadcasting Company.

History did not record what FDR thought of his appearance on those first television sets or if he even saw himself on the screen. But suppose he did see and suppose he hated it. Suppose he felt that the folks at NBC were out to make him look bad. Suppose he was so mad that he wanted to force the network to apologize and change their ways.

He would have been screwed.

He could have complained, but if he wanted those famous fireside chats to reach tens of millions of Americans via NBC radio, he could never boycott the network.

Donald Trump is no FDR. And NBC no longer enjoys a monopoly. For the past two weeks, the Republican presidential front-runner has refused to grace our broadcast airwaves. He is fuming about my tweets from Raleigh, specifically the one where I said he left the stage "abruptly" after protesters interrupted him. The word *abruptly* really pissed him off. Now Trump wants me to apologize for my "dishonest" reporting. Problem is, my reporting was accurate and journalists do not apologize for accurate reporting.

This is a big fucking face-off.

NBC doesn't have a solo hold on people's attention. We compete with CBS and ABC, CNN and Fox News, plus Twitter, Facebook, Snapchat, YouTube, and the millions of people on this earth who own a smartphone or computer attached to the World Wide Web. They all have the personal power to broadcast. That's why I'm sitting in an office in 30 Rock, thinking about the old RCA Building and waiting for Donald Trump to call me.

Apparently tweeting that I should be fired, calling me a liar in front of millions of people on national television, and receiving death threats from his followers shortly thereafter was not enough punishment. He wants penance. He wants groveling. He wants to hear those two precious words. And until he gets them, he says, no NBC News. No *Meet the Press*. No *Today* show. No *Nightly News*.

The phone rings.

"Katy, it's Donald."

He actually sounds a bit friendly, making small talk about his poll numbers. He wants my opinion of Ted Cruz and Ben Carson. He is talking to me as if we're old friends, and it occurs to me that in his mind maybe we are. At least for the moment. Banter is part of his process. He's a person who crowdsources. He likes to get everyone's take. He'll call anyone who will listen—friends, loved ones, business partners, lawyers, rivals, and, yes, even reporters. He was famous for it in the New York tabloids—calling to hear opinion, spread gossip, or just hype himself.

"Trump bought reporters, from morning paper to *Nightly News*, with two minutes of purring over the phone," Jimmy Breslin wrote in *Newsday* in 1990. I'm invoking Breslin again, but he was one of the reporters who saw Trump for what he was before anyone else. One Queens guy on another. When Trump was dangling over the edge of a financial cliff—the grip of his pinky finger the only thing between him and a reported nine-hundred-million-dollar bankruptcy, Breslin correctly predicted his comeback. All Trump had to do was boast.

"He uses the reporters to create a razzle dazzle: there are five stories in the . . . morning papers leading into 11 minutes of television at night," Breslin wrote.

The financial people, who lead such dreary lives, believe what they read and see on television. Trump is larger than life. No, not Trump. Don't use that name. It's Donald! He cannot lose. . . .

Trump will call and announce his rise. The suckers will write about a heroic indomitable spirit. Redemption makes an even better tale. So many bankers will grab his arm the sleeve will rip. All Trump has to do is stick to the rules on which he was raised by his father in the County of Queens:

Never use your own money. Steal a good idea and say it's your own. Do anything to get publicity. Remember that everybody can be bought.

And that is exactly what he did. He's spent the last thirty years bragging about his gut, his brains, and his success. He's still doing it. He is the living embodiment of the old maxim that if you say something often enough, people will believe it. That's why I'm even having this conversation. Because despite what the insiders might say, Trump isn't going anywhere.

That's also what worries me. After about a week without exclusive Trump interviews on our airwaves, I apologized to the executive producer of the *Today* show, since Trump was now making more appearances on *Good Morning America*. To his credit, Don Nash said he could not care less. Don't apologize, he told me.

To have support from the company is everything—but I still want to smooth things over, because this is an obstacle and I want to get back to reporting.

"Well, I know that you're busy, so let's just get into this," I say. "I know that you have been less than pleased with what you've seen as some unfair reporting."

I don't have an office, so I'm huddled inside Dafna Linzer's space. She's a veteran reporter and, for a call like this, both my editorial backup and my witness. We don't trust that Trump will portray the conversation accurately. If he decides to bring it up later, as we could easily imagine him doing at a rally or on Twitter, we wanted two pairs of ears and two sets of notes for the record. Dafna and I trade glances and I carry on.

Intense scrutiny is what comes with being the front-runner, I remind him. The American public deserves to know as much about their potential president as they can. Trump listens, but he still has complaints. He didn't like the way I characterized his departure from the stage in Raleigh. He had laryngitis, which, okay, he did. And he says that protesters did not force him off the stage, which, hold on, I never said.

This is really about image. Trump cannot bear looking weak. His whole pitch to the American people is "strength and stamina." He's the outsider who is willing to say what the others won't, to do what the others are afraid of doing, to fight for you. He is a man who cannot be intimidated. This obsession with old-fashioned power is why he's so enamored of Vladimir Putin, who rides horses bareback and shirtless.

Something seems to click with Trump. I don't know what. But he seems to accept my explanation of my job, to take the scrutiny as a sign of respect. Maybe it was enough for him to hear me say that he's got a serious shot at the presidency. He's suddenly ready to move on.

"I appreciate what you're saying," he says. "Take care of yourself. . . . Be fair to me, Katy. . . . You and I should be friends."

We hang up.

I did not apologize.

Everybody that goes against me, it's like, X, X," Trump says, on December 21, 2015, air-drawing the Xs with his finger. "Wouldn't it be nice—that should happen with our country. Everybody goes against us—down the tubes."

The crowd roars here in Grand Rapids, Michigan. Trump's rhetoric is getting scary. He's talking about rivals dropping out of the presidential race, an imaginary X appearing next to their names. But it's not clear what he's talking about when he says, "Everybody goes against us—down the tubes."

Who would go down the tubes? Critics? The opposition party? Non-Trump voters? And who would send them down the tubes? Who is this *us*? Trump staffers? A legion of yes-men? What does he mean by *down the tubes*? Professional failure? National expulsion? Public embarrassment? Death?

Trump would deny the menace in these metaphors, but there's a calculated vagueness to them. Some supporters think he's joking, while others feel he's laying the foundation for a militia they'd like to join.

He set the tone in his announcement speech, warning that America was getting "weaker." Now his message is all about America getting tougher and stronger and, in winks and nods, more violent and unforgiving, too.

At a town hall in New Hampshire in late August, Trump brought up the case of Bowe Bergdahl. The troubled young army sergeant had walked away from his post in Afghanistan, setting off a dangerous search and ultimately a prisoner swap with the Taliban. Trump had a different idea.

"Bing, bong," he said, pretending to hold a rifle and squeeze the trigger as he did. That idea—that Bergdahl "should have been executed," as Trump said in October—continued to surface and resurface at Trump's rallies, even as he introduced new visions of violence.

In late November he imagined an America that waterboards its enemies, on the logic that even if the simulated drowning doesn't work as an interrogation technique, it works as a punishment. "They deserve it," he said at a rally in Ohio.

Earlier this month he invited a crowd to imagine an America that pursues terrorists with an indiscriminate rage, unconcerned if the missions end up killing the terrorist's wife or children.

"They're not so innocent," Trump told me.

A few weeks ago, he turned his attention to enemies within. At a rally in Birmingham, Alabama, a black protester—and well-known local activist—disrupted Trump's remarks. A fight broke out and security moved in, hauling the protester toward the exits. But the protester fell on the way to the door and a group of white men closed in on him. He appeared to be kicked and punched, according to video. A witness also described him being choked. Trump didn't condone the violence on the spot, but the next day he said the guy was "so obnoxious and so loud" that "maybe he should have been roughed up."

Journalists who cover Trump without being in the room will sometimes say that Trump's crowd isn't with him. But I can tell you, the crowd loves it. There is no rush for the exits, no howl of disgust. The first rally in the aftermath of the scuffle in Birmingham was as packed as the last—maybe more packed.

People seem drawn to Trump's rallies in the same way that they are drawn to a professional wrestling match, and as with a professional wrestling match, they seem divided between people who believe all they see and hear, and those who know it's partially a performance. The scariest thing about being at a Trump rally is that you don't know who believes it and who doesn't.

I pick up a brown bag next to me on the press riser. Trump lapses into his same old stump speech and I'm about to lapse into knitting—yes, knitting. I have four half-knit scarves tucked away in a drawer or box somewhere. I keep starting one only to get bored. But this time, I swear, I'm going to finish this goddamn scarf. I want to give it to Ali Vitali as a Christmas present.

I've promised the embeds I would make one for each of them. I'm still the only regular correspondent on the Trump beat, so usually it's just me and them on our daily flights to nowhere. They're all twenty-something. Sometimes I can feel like their mother—patiently listening to all the reasons why they are not getting the respect they deserve, or scolding them when they act out and stick their entire grubby hand in my salad.

Don't ask.

They make me feel better about my age. Your twenties are a dark tunnel of insecurity and frustration. You know so much more than anyone gives you credit for. You're too good for logging and shooting cuts. You should be the one in front of the camera. Believe me, I hear you.

But sometimes I want to grab Ali by the arms and scream about everything I had to go through to get to this point. I want to tell her about the fight to be taken seriously. And I want her to open her eyes and look around, because she's on the biggest story of the century. But I don't. She's a good kid and a smart kid. She'll get there. It isn't until about year ten that you finally realize how little you actually knew when you started and, like a drunk who just got sober, you want to call all your friends to say, "I'm sorry I was such a shit."

I'm rushing to finish her scarf now, so I can give it to her before I leave for London tomorrow. I know I look ridiculous. I'm surprised no one has tweeted a picture of it yet, claiming that I'm not paying attention.

But there's no ignoring Trump.

They said, you know, he's killed reporters," Trump says from the stage.

I put down my needles.

Trump is talking about Vladimir Putin again. As Trump flirts with violence and authoritarianism he seems to avoid any criticism of the Russian strongman.

Last Friday on *Morning Joe*, Joe Scarborough reminded him—incredulously—that Vladimir Putin "kills journalists, political opponents, and invades countries." Trump was on the phone, his voice raspy. He sounded like he was calling in from bed, and he didn't care a bit about Joe's incredulity.

"He's running his country, and at least he's a leader, unlike what we have in this country," Trump said.

"Again," Scarborough said, slowing down the conversation, "he kills journalists that don't agree with him."

"Well, I think our country does plenty of killing also, Joe," Trump responded. He sounded like he was shrugging his shoulders, shrugging off America's moral high ground. "There's a lot of stupidity going on in the world right now, Joe. A lot of killing. A lot of stupidity, and that's the way it is."

No one understands Trump's infatuation with Putin. He's been talking about him for years, sometimes praising him at odd times. For instance, back in 2007, after Russia was accused of targeting Estonia's government, media, and banks in a massive cyberattack, Trump gushed about how the country was getting back on track. "Look at Putin," Trump said. "He's doing a great job in rebuilding the image of Russia and also rebuilding Russia period."

In 2013, right before he held his beauty pageant in Moscow and right after Putin passed a law against gay "propaganda," Trump tweeted, "Do you think Putin will be going to The Miss Universe Pageant in November in Moscow—if so, will he become my new best friend?"

Putin was jailing homosexuals and his government was accused of turning a blind eye to the roving gangs brutally beating gays and lesbians. The Western world was horrified and Trump, a sixty-seven-year-old man and future presidential candidate, was tweeting Putin a love letter.

Now, in this Christmas-themed arena, Trump is mulling the idea of following Putin's lead when it comes to journalists. Killing journalists? At first he waves it off.

"I'd never kill them," he says. He pauses and smirks.

"I hate them, but I would never kill them. I'd never do that."

He pauses again and makes a so-so gesture with his hand, as if entertaining a bloodbath in the press pen, as if truly considering whether there could be any circumstances where he might change his mind and decide that one of us back here actually needs to die.

"No, I wouldn't," he finally says. "But I do hate them."

Here we go.

"Some of them are such lying, disgusting people," Trump says.

So much for Trump and me being "friends."

The crowd loves it. They turn as one to boo at us in unison. Six thousand Trump supporters railing against thirty or so journalists—caged in the center of the arena like a modern-day Roman Colosseum.

When do they release the lions?

What happens if someone in here can't take a joke?

Trump doesn't seem to care. He points to the still pho-tographers who have gathered at his feet to take pictures and snarks about how they've just been let out of the "cage."

This is not normal. I've been nervous for the past two weeks, but now I'm just angry. What if someone gets hurt? What happens then? Do people think we enjoy living our lives on the road, dragging our suitcases behind him all over the country? While Trump has his private plane with a bedroom and all the fast food he wants, we have faceless airports, cramped coach seats, and peanuts. And while he's zipping back to Trump Tower to sleep in his own bed, every night we're in yet another random, far-flung hotel, hoping our loved ones are up when we finally make it to our rooms, exhausted and grumpy. This job is hell. On relationships. On your body. On your mind.

We do it because it is important to show the public who is running for president. It's important to show how they be-have. How they think. What they believe. Who they admire and why. Yes, we give Trump a ton of airtime and article space. But that's because he is unlike anything anyone has ever seen. And despite what folks who don't like him might want to argue, he is resonating. And we have an obligation to document it.

The crowd roars and laughs. I scan their faces. I'm look-ing for one woman in particular, a hairdresser I met in the bathroom here a couple of hours ago, before the crowd cheered my demise. It was a one-stall public restroom. Concrete floor. Plastic trash can. The vanity mirrors were not glass but a thin sheet of metal on a cinder block wall. The aesthetic could only be described as Soviet.

Fitting.

All day I had been trying to get reaction from other Republicans, but in the bathroom I had a more immediate goal: my hair. I barely slept last night, and this morning I didn't have the energy to preen. I got by with a comb earlier when we were outside doing live shots. But the wind and the water eventually ruined me, so about an hour before my *Nightly News* live shot I was at war with a curling iron.

Being a woman is a pain in the ass. You have to look "good." Your hair needs to be neat—not just combed through, but "done." Blow-dried, ironed, curled, sprayed. Your face needs to be enhanced. Foundation, powder, eye shadow, mascara, lipstick, blush, contour. Your clothes have to look sharp, too. And you can never wear the same thing twice—at least not in the same week. A guy can throw on the same suit every single day for a year and no one would notice. I'm not exaggerating. An Australian broadcaster tested it out. His coanchor, a woman, kept getting letters, e-mails, and tweets from viewers criticizing what she was wearing. He was appalled. He never got notes. So he wore the same blue suit day in and day out. Three hundred sixty-five days. Surely someone would complain. No one did.

"No one has noticed," he said at the time. "No one gives a shit."

I figured out a work-around to the clothes, at least for the winter. I bought the same J.Crew sweater in fifteen different colors: Neon Primrose, Misty Peri, Light Citron, Navy. That, along with a rainbow collection of scarves, can trick any viewer (or boss) into thinking I travel with a cart of suitcases.

Still, I longed for that kind of blessed apathy as I tried to plug in the iron, which in itself was a trial. I had to snake the cord around the trash can, and even then the curling iron didn't make it all the way to the mirror. I made up the difference with my neck. All the while female Trump supporters streamed in and out while Trump's rally soundtrack bounced off the concrete. Everyone looked at me like I had lost my goddamn mind, and the truth is, I was close to it. I could not wait until the sun set on December 21. I missed my friends, my clothes, my life in London. Most of all, I missed my bed, which I'd slept in four or five nights since June. But starting tomorrow, for ten full glorious days, I would get to go home to London. I would get to soak in everything I missed.

That's when I first saw her.

"Want some help?" she asked.

I didn't immediately answer. The curling iron was red-hot, this was a Trump rally, and I was only a couple of weeks out from having essentially been named public enemy number one by Trump himself.

Does she look crazy? Do I?

Sensing my hesitation, she laughed and said she is a hairdresser in town. Something about her smile made me trust her. Besides, with her help I would get out of this bathroom sooner.

"I'd love some help, thank you," I said, handing her the iron.

We made small talk about the absurdity of the situation. I told her about my long days and how hard it is to look presentable on national TV when you're living on the road. She talked about the upcoming holiday and how excited she is to see Trump in person. She was lovely in every sense of the word, and she was a good reminder that Trump supporters are more complex than the one-note candidate they cheer onstage.

I left that bathroom feeling lighter.

Now I feel weighed down by the whole scene. Four days before Christmas, the stage set with poinsettias and lighted wreaths, supporters wearing holiday sweaters—and the crowd is cheering about the idea of killing journalists. I know I shouldn't take it personally. I'm a professional. But to the lady who curled my hair in the bathroom, who is now somewhere in a crowd that is laughing at the idea of Trump killing me: Thanks, my hair looks great.

Has anyone on earth ever wished for a longer overnight flight? I did. I could've spent another four or five hours sleeping, watching movies, and drinking, blissfully free of the Internet and my phone. Yes, there was a tug at my gut. A worry that I might miss something while I was in the air. Something big, like Trump dying. Don't make a face. This is a universal worry for reporters. Your candidate dying is the one story you absolutely cannot miss. If you don't believe me, read Timothy Crouse's *The Boys on the Bus*, the book about reporters covering the presidential race in 1972.

Crouse writes about it on page 2. This is why you see me, Hallie Jackson, Kristen Welker, Kasie Hunt, and Andrea Mitchell on TV from 6 A.M. to 10 P.M. This is why we can't have a fluid conversation off camera without checking our Twitter, texts, or e-mail. We are addicted to the "what if?"

I land in London at 9 A.M., a little groggy but thrilled to be home. I practically run to customs and immigration and swell with pride when I get to use the registered traveler line. Outside it's cold and gray. I can smell the rain before I feel it, but even though I'm dragging two giant suitcases onto the train into central London, I don't mind. Paddington Station is packed. I take a taxi the rest of the way. The cabbie substitutes *f*'s for his *t*'s when I ask him how he is doing. "I'm all right, fanks." The car is driving on the wrong side of the road. But I am finally on the right side of the pond.

My flat is a time capsule. Thanks to a friend, there's no rotting milk in the fridge or stale clothes in the dryer. I lie down and close my eyes. It's good to feel my own sheets. To bury my face in my own pillow. To sleep. At home. I nap for an hour or two, and even though I'm thirty-five hundred miles away I get up with New York and instinctively grab my phone to see what's happening. I start to tell myself not to care, but then I see it. The e-mail I knew would come, asking the one question I do not want to answer.

Can you pack up your flat and move back to New York?

I drop my phone on the bed, roll over, and start crying.

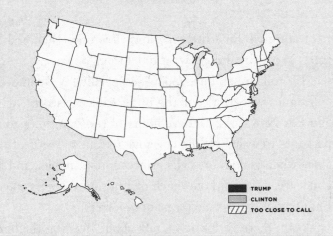

NBC NEWS HEADQUARTERS

MIDTOWN, NEW YORK CITY
12 P.M.–3 P.M., Election Day

W elcome to the 2 P.M. hour of MSNBC live. It's Election Day and I'm Katy Tur." I'm anchoring. The call
came a couple of days ago. How'd you like to take the helm
at two on Tuesday? Janelle asked. I could tell she was smiling. She knew I would be thrilled. She knew I wanted it. She
knew, I knew, it was a big deal to anchor Election Day—a
big vote of confidence.

I pored over scripts, made last-minute adjustments, and
right before the camera came to me, I took a long deep breath.

*Just read. And if you somehow forget how to read, vamp.
Who knows this better than you? No one.*

I admit, I am a little nervous. I've only anchored five times in my life. I don't want to let them down.

TV folk are all busy today—especially my Trump-beat colleagues who work for the other cable networks: CNN and Fox News. We all have hours of programming to fill, packages to write, rolling coverage to follow. For us, this is a day when you ignore calls, cancel appointments, and find childcare, a day without much of a lull once it gets started. But none of this is true for print reporters.

Yeah, there's a lot of big talk about "feeding the beast" online. But the Trump beast is twiddling its thumbs.

So Ashley Parker of the *New York Times* is eating soup dumplings alone at Joe's Shanghai, down the street from the Hilton. Ben Jacobs of the *Guardian* just woke up and *Politico*'s Eli Stokols is hunting for a tie at Barneys. The *Washington Post*'s Jenna Johnson isn't even in New York City. She's in Pennsylvania, at some farmhouse turned into a Trump shrine, writing what's likely to be a poignant story of Trump supporters paying homage to a doomed candidate.

Other reporters at their newspapers are busy. But the Trump campaign reporters aren't, for a simple reason: they've planned ahead. Jenna is still working on that farmhouse piece—the farmhouse, by the way, is painted like an American flag and guarded by a fourteen-foot-tall metal Trump—but the others have the broad outlines of their postelection stories already done. Ashley filed a "last scene" piece for the *Times* on the press plane at three this morning. Ben, who has been on the beat for the *Guardian* at least as long as I have, already has his "postmortem" interviews lined up on how and why Trump lost. Likewise, Eli has been working for days on his version of the Trump loss. He is basically finished with it. All that's left to add is a little color about what happens after the polls close. It's not that the media is trying to prejudge the news or force an outcome. It's that speed matters, and certain events you just see coming. Ever wonder how we get a sharply produced obituary of a celebrity or political figure on television minutes after we find out they die? That's right—we prewrite it. TV is a macabre industry. We have obituary reels already written, voiced, and edited for all the older movie stars or world leaders, and a few for the ones who just can't seem to stay on the wagon. I'll let you take a guess at whom. Sometimes, you can count the chapters of a significant life by the number of names on the byline. By this Election Day, sixteen *New York Times* reporters had worked on Fidel Castro's obit. Sixteen! The first writer started in 1959. Life is unpredictable. Most times, so is death.

The point is, you prepare when you can. And in this case, the polls are so unbelievably lopsided, you prepare for Trump to lose.

So Ashley finishes her dumplings and orders a hot and sour soup. When she's done she walks a block to Trump Tower and meets her colleague Maggie Haberman. They pass the time at the Starbucks in the lobby. Trump aides are carting carafes of coffee up the elevators to the election headquarters. The early returns don't look good. So they write a color piece about what the end is like in the place where it all began.

Eli buys a blue tie with thin pink stripes, then goes for a walk in Central Park. The leaves are changing: bloodred, copper, and gold. He takes a picture of the Bow Bridge and posts it to Instagram. Caption: "Pre–Election Night serenity walk in the park." It is not lost on him that fall is when everything dies. Ben gets up, but he doesn't go far. He hangs around the press hotel. The *Guardian* is doing a live blog about what the election will mean for women. But he has a funny thought. He calls his editor and suggests something crazy: Trump could win.

6

"Find That Asshole Tur!"

OCTOBER 25, 1983

Thirty-Three Years and Two Weeks Until Election Day

My mom likes to say I've been covering breaking news since the day I was born—longer, if you count my time in utero. The day she went into labor, she and my dad were in Hollywood, covering a shooting. A fire department source of theirs said they wouldn't want to miss it. Note: when a source says, "I can't tell you why you should go, but I promise it will be worth it"—you go. In my mom's case, you go, even when it means lugging a seventeen-pound tape deck and you're nine days overdue with your first child.

The story involved Jerry Dunphy, the lead anchor of KABC's 11 P.M. newscast, famous for his sign-on: "From the desert to the sea to all of Southern California." He'd been shot in a mugging gone bad, just an hour before his show, in the parking lot of KABC. My parents got the tip and left the house in as much of a rush as they could manage. There was never a question of whether my pregnant mother would join my father on the hunt. They were a husband-and-wife reporting team, proprietors of the Los Angeles News Service, and my mother wasn't the kind of woman who, in her words, "sits around the house all day waiting to give birth."

Besides, they both needed to be there. Video cameras in those days didn't record tape on their own. They were attached by a cord to a tape deck, which meant all camera crews had to have at least two people: the camera operator and the person holding the deck. In my parents' case, Bob held the camera and Marika handled the deck. When they arrived at the Prospect Studios in Los Feliz, they were the only journalists on the scene. That was a reason to speed up, not slow down. My father rushed ahead, looking for the best shots, all but dragging my mother and by extension me, all of us attached by one cord or another.

The gambit paid off. ABC's anchorman had been taken to the hospital, but the police were still on the scene. My parents got such great footage of the investigation that they sold it back to KABC itself. Afterward, they settled down for a late-night banquet of Chinese food. That's when the contractions started. Eighteen hours later, the Tur family had a new member of their news team: Katharine Bear Tur.

Tur is my father's last name.

Katharine is for Katharine Hepburn, a woman who wore pants when everyone else was wearing skirts.

Bear because—well, never mind why *Bear*. This book

only has so many pages, and I'm a journalist, not a licensed shrink.

By the time I was born, in 1983, my parents already had a history of turning traditional family milestones into reporting trips. They pursued their very first story on their first date. The date itself was a big get for my father. He was eighteen, pretending to be twenty-one, and my mother was twenty-three, pretending not to enjoy the attention. She was working the ticket window at a movie theater near UCLA, where she was a graduate student in philosophy. He was—well, it wasn't clear what he was at the time.

"He used to walk by the box office with his camera every night. Just a kid on a school assignment, I decided. He would pretend that he was taking my picture and I would kid him about sending it to the National Enquirer," my mother wrote in her diary back then. She was flattered but not immediately interested. Bob Tur persisted. He asked her out every day for a month until she finally agreed.

It was Halloween 1978. My dad decided to impress my mom by taking her up for a ride in a Piper Archer, a one-engine airplane he was learning to fly on a lark. She hated the flying, but she liked the cocky young man at the controls. Back on the ground they decided to extend the date—in a manner all their own. My dad had heard about a school nearby that had been vandalized. He proposed they go take some pictures, try to sell them to the *Los Angeles Times*. They took the pictures and again they wanted to extend the date. This time my father proposed they drive downtown to Skid Row. Someone had been stabbing homeless men to death. If they could spot the guy and snap some pictures, it would make their careers in a single evening.

Skid Row was their second date as well. And their third. For weeks, while a typical young couple might have done dinner and movies, my parents pursued the Skid Row Stabber. They never found him, but in their youthful fever for journalism they discovered a purpose for two lives that were otherwise adrift.

My mother had grown up happy, if a little bored, in the working-class section of Beverly Hills, south of Olympic Boulevard. Her mom ran the house. Her dad built pools for Bel Air's rich and famous. She was the middle child, sandwiched between an older brother and a younger sister. By college she wanted to break out a little, be a painter or a writer, the next Claude Monet or Thomas Mann. As she toggled between these ambitions, she said yes to a Ph.D. program in philosophy at UCLA and took a part-time job selling tickets at a movie theater—because it gave her time to read during shows. At the same time, she fell for journalism in the afterglow of Watergate.

My father was a harder case, a teenage runaway with a big mouth and bigger ambitions. His own father had been a gambler, a bad one, the kind whose family could never really be sure whether they'd go to bed in the same place they had woken up. My dad came home from school many times to find that his family was gone, forced to move somewhere new, one step ahead of the eviction papers. He would sit on the curb until his father picked him up, the old man often angry and resentful of the chore. Jack Tur's ego couldn't bear the gambling losses that were ruining his family. Instead he blamed his kids.

Because my dad was the oldest, he took most of the blows. Because my dad was the oldest, he got away first. He left home at fifteen. The day he spotted my mother in that ticket booth, he was working as a licensed paramedic (don't ask me how), spending his nights in the very same ambulance he used during the day. He had chutzpah and charisma but no cash. She had stability and reason but no adventure. They were a perfect team.

They started Los Angeles News Service as a stepping-stone operation, a way to get LAPD press passes, on their way to what they imagined would be big-time network jobs. But those jobs require you to stand in one place and shovel journalistic shit, aka wait in line and pay your dues. It wasn't for them.

Instead they turned LANS into a full-fledged operation, less a stepping-stone than a middle finger to the competition. They bought a fancy street map and an emergency scanner. They internalized the police and fire codes, all the lingo, all the neighborhoods. And they never dialed down the scanner. In my first years of life, it was my music box and my bedtime story. It's the reason why my first word was *hot*, my mom says, only somewhat joking, and why my second and third words were *smoke* and *showing*.

My parents made their way on wits, guts, and a creative interpretation of fair game. To make their original jump from print photos to video, for example, they needed seventy thousand dollars for a camera. They raised it by bending their journalistic skills to the task of private investigative work. Yes, they were licensed PIs—except that didn't pay well enough, fast enough. So they turned to another twisted sister of journalism, the growing field of bounty hunting. If you captured a fugitive for Los Angeles County in those days, and you were a wily negotiator like my dad, you got 20 percent of the bond as a bounty. That's how my parents got a down payment for a news camera—one fugitive at a time.

Ingenuity also got my parents in trouble. In 1985 they were ready for their first news helicopter, sticker price: $250,000. At the same time, all the news crews at KTLA were on strike. Maybe they had good grievances, maybe not. Either way, my parents ignored the strike and worked. They had a one-year-old at home, another kid on the way, and a burning need to get ahead. Some in the field thought they did a dirty thing, and a lot of crews have never forgiven them for it.

But that was only the first part of a two-part play for the news helicopter. After a few months with KTLA, my parents had squirreled away thirty thousand dollars. With that in the bank, my dad walked into the sales office of the Hughes Helicopter Company: a twenty-four-year-old asking for a quarter-million-dollar flying machine with 15 percent down. They laughed him out of the office. So he walked over to Bell Helicopters with a thicket of invoices from KTLA and a plan to run a freelance news-gathering operation from the sky.

The salesman bit, and TV news would never be the same.

Los Angeles News Service was not the first to use a helicopter. But Bob and Marika Tur were the first to do something memorable with one. They would cover fires, shootings, and, most unforgettably, police pursuits. Their first big get was Madonna's 1985 wedding to Sean Penn. Bob didn't yet have his license to fly, but he could sure ride shotgun, call the FAA for clearance, and direct a hired pilot to pull this new helicopter a mere 150 feet from the bluff where the stars were wed. Madonna flipped Bob Tur the finger. He sold the stills of her doing it for six figures.

By 1987, my father and mother settled into a new routine: him flying, her on the camera. She was fearless, hanging out over the skids with a thirty-pound machine on her shoulder. She was also shameless. And I mean that as a compliment. With no way to send live video in those days, my parents would fly tapes from station to station, dropping them from the copter down to the roof where some producer would be waiting. To keep them from breaking on impact, my mom wrapped them in anything she had on hand. Usually, that meant her clothing. On a busy news day it wasn't unusual for her to get back to the hangar in only her underwear.

In addition to flying, my father filed live radio reports, his hands on the stick and his eyes on the news unfolding below him. He was new to it, and he was practicing, which might be why he also started asking me, at age four or five, to work up my own live reports. He'd point the camera at me and cue me. In one of the surviving tapes, I tell the story of an imaginary fire in San Diego that ended with all my friends and me having a party at McDonald's. And you know what? I wasn't so bad. I didn't have a nanny. My parents didn't have the patience for preschool paperwork. My entire early childhood education was tagging along with them—witnessing car accidents, multiple-alarm fires, and shoot-outs.

Looking back now, after more than a decade in live television myself, I appreciate my father's flaws as a reporter—even more now because I see them in myself. His delivery, like mine today, was filled with "uhhs" and "umms." Sometimes, like me, he'd just blank. During a massive flood in the agricultural region of Ventura, he couldn't remember the word "lettuce." "I'm flying over the salad fields," he said on live TV. *Salad fields?* No, he wasn't perfect. But he was real. And when paired with my mother's pictures, viewers connected with the product.

Their big television break came in 1991, not long after they upgraded again, adding a microwave dish—for live video—to the underside of their copter. That February a 737 collided with a commuter plane at LAX and erupted in flames. The FAA closed the airspace to all aircraft, but the L.A. Fire Department called in an exception—for my parents. It wasn't a favor. My parents had a thirty-million-candlepower "Night Sun," a fancy term for a very bright spotlight. The LAFD wanted it. And my parents got their exclusive.

The following year brought the first police pursuit ever covered live from the air, a no-joke watershed moment in media history. It started with a carjacking murder, which turned into a high-speed chase through the city's maze of a freeway system. Bob and Marika Tur were overhead. At the time they had a contract with a station airing a rerun of *Matlock*, which left the news director with a decision: stay with *Matlock* or switch to the red Cabriolet.

He took the car chase. My dad narrated as my mom tried to keep the camera trained on the driver. For the next forty-five minutes, the driver sped on and off the freeway, wove through busy city streets, and periodically pointed his shotgun out the back window to fire on the cops. The chase ended with an empty gas tank and a hail of bullets. The cops shot the suspect to death on live TV. And my parents had it exclusively for twenty-five minutes before any of the other stations got their helicopters to the scene. That's a lifetime in TV news.

The next morning the ratings came out. The news won. The city was high on the adrenaline of the chase, and my parents were the pushers. But like any addiction, you don't know how much damage you've done until it's too late. Bob and Marika covered 229 live pursuits after that. Today, their former colleagues blame them for the downfall of local TV news. Some would say the downfall of national TV news, too. They don't dispute it. I don't, either.

The high couldn't last, of course. It never does. For a while I loved what my parents did for a living. How could I not?

As an elementary school kid I spent my afternoons and weekends flying over Los Angeles, where I developed an obsession with backyard pools. *Everyone in L.A. has one, why can't we?!* When the news was slow, my dad would fly us to Catalina Island or Santa Barbara for lunch. Or we'd buzz the beach and wave at the sunbathers. One of my earliest memories is of my dad doing live radio reports of the Rose Parade. I was already so comfortable in the air that I un-buckled my seat belt, got up, and opened the chopper's side door so I could get a better look at the flowery floats below. I was holding on to the door's handle and looking down when my dad calmly turned his head and asked me to sit down. He then reached back to close the door, the helicopter swaying ever so slightly. He told me later that he'd had to land im-mediately because he nearly had a heart attack. I wouldn't have known. He was that cool under pressure.

Kids are kids, though, and before long I thought the cockpit Christmas cards with me, my brother, Jamie, and my dog, Daisy, wearing flight headphones were "boring." I thought getting picked up at sleepaway camp in a helicop-ter was "so annoying." And I felt that flybys of my summer softball games with my dad cheering me on over the heli-copter's loudspeaker were "mortifying." The novelty of the helicopter had worn off, and by middle school I was tired of being asked to do a live report every time I got in the car. I decided I wanted nothing to do with the news.

I maintained that point of view even when a truck driver named Reginald Denny stopped in the wrong intersection at the wrong time at the outbreak of the L.A. riots. My mom and dad were on the scene. When Denny got pulled from the driver's seat by a group of gang members, my parents got lower. When the mob had him on the ground and were kicking him in the back and head, my parents got lower and lower again—as low as seventy feet. It wasn't about the shot anymore; it was about trying to save the man's life by scaring off the crowd. But the crowd closed in and the cops were nowhere in sight. My dad was outraged. With millions tuned in, he declared that the LAPD had abandoned the city.

Back on the ground, they say they saw bullet damage in the engine blades. A huge deal, since the engine could've stopped midair. My mom says there was even a bullet hole in the camera battery stowed beneath her seat. It was a miracle she wasn't shot as she pulled in the footage. My brother and I watched it all from our grandparents' house. Then we lived through the aftermath, not only of the riot—but of the journalism. Gangs were angry about my parents' coverage. We started getting death threats. It wasn't safe for us kids to be home without our parents. And my father got his first concealed-weapons permit. For years he wore a gun on his belt every day and slept with it under his pillow.

Not even O. J. Simpson's strange, slow-speed chase was enough to reignite my interest in my parents' profession. In fact, when I saw their helicopter hovering over my school one day, I was convinced they were trying to spy on me and my friends. They were not. O. J. Simpson had led cops to his house, which happened to be nearby. Once again, my parents in their helicopter were the first to spot him, ten minutes before the competition. My father remembers hearing rival assignment editors screaming over the radio, "Find that asshole Tur!"

What he's less likely to recall is how much he really had become an asshole. He and my mother were an Emmy, Peabody, and Murrow award-winning team, with more than a decade of experience in L.A. news. But they never left their hangar, never mixed it up in the newsroom, considered everyone to be the dreaded competition. That left them little room for friends and no room for alliances. All people really knew about them was what they saw on TV or what they heard over the squawk box. It was ugly.

My dad had a hot temper, which the highly pressurized cockpit didn't help. To him, Mom could never hang out of the helicopter far enough, focus on the right thing, or move fast enough for him. God forbid she "double-clicked": accidentally pressed the record button twice, meaning she only found out she wasn't rolling until after she was done. It wasn't that he didn't trust her. He never yelled during a big breaking story. The stress overtook him when he had the time to think. Later on my mom would admit that, yes, sometimes she did it on purpose, but at the time she always denied it, further fueling his rage.

My grandmother Judy was his only buffer. She could talk him down—quieting his belief that the world was out to get him. She was twenty when she had my dad, forty-three when he had me. She was so young that she refused to let me call her Grandma, which is fair enough. She had shortcomings as a mother her first time around, mainly her failure to stand up to an abusive husband. But she was a wonderful second mother to me and my brother. She picked us up from school, made us dinner, and consoled us when the other kids were mean.

She held everything together, including my parents' business. Judy and I would drive down to Hollywood or Burbank, make the rounds. I knew all the city's newsrooms as a kid. We were L.A. News Service's best asset: a cute kid and a lovable grandma, warm and disarming. She built relationships with the city's assignment editors and news directors. She was also shrewd with footage. At one point she was flipping video of the Reginald Denny beating for five grand a use—helping to turn my parents into on-paper millionaires. At least for a little while. We had a seven-figure helicopter, two Porsches, and enough extra cash for biannual ski vacations and a trip or two to Hawaii.

Then, in 1998, everything fell apart. My dad had heart surgery. Judy complained about a pain in her toe. KCBS stopped working with my parents—just after they had bought a $1.3 million helicopter. Soon my parents couldn't make the payments. Meanwhile, the pain in Judy's foot wasn't going away and now she was coughing. She went to the hospital thinking it was emphysema. She left with a diagnosis of lung cancer. The details here are murky. I was fourteen, and I didn't know my entire life was about to change.

In the span of three months, my grandmother died and I lost the only parents I ever knew. I don't mean that they died. I mean that Bob and Marika were never the same after they lost Judy and then the helicopter. The business was their identity. Without it, they didn't know what to do with themselves. Instead of looking to find a new path, something they were always so good at, they froze like a watch in an explosion. There was no more future, only the past. So they lost themselves in copyright lawsuits, suing KCBS, KNBC, NBC News, Reuters, and anyone else who aired their video footage without first getting a contract.

It made them enough money to survive, but the bills piled up. Rent was often late. We stopped answering the phone because it was always a bill collector. Instead of going east for college, I went to a state school, UC Santa Barbara, and took out student loans for everything. To this day, the last semester of my brother's private high school tuition is still unpaid.

The worst part about it wasn't losing the money. UCSB is a great school. Paying off student debt makes you a better person. The worst part was that I lost my role models. They were depressed and angry. Everyone in the business hated them for their lawsuits. All they seemed to do was fight. It was toxic. The news ripped my family apart.

I went to college with my sights set on becoming a doctor or a lawyer. Stable. Predictable. Then a funny thing happened. It was my senior year and I was driving back to Santa Barbara from Los Angeles with my college boyfriend. There was a roadblock in Malibu—another brush fire. Not a big deal. But I found myself pulled toward it. Long after I'd stopped expressing an interest, my dad kept forging me press passes and forcing me to carry them with me. The one I had in my purse that day was my grandmother's. My dad had pasted my picture over hers and relaminated it. *What the hell*, I thought. "Let's see if we can get in," I said to my boyfriend. I pulled up by the officer guarding the road and flashed my pass.

"Who do you work for?" he asked suspiciously.

"Los Angeles News Service," I said.

He looked down, then back at me, then down again.

"Where's your gear?"

"My crew is up ahead. They have the cameras."

"All right," he said. "Be careful."

To me it was pretty clear he knew I was full of it. I was twenty years old. But my boyfriend was awestruck. "I've never seen you more confident than you were just then, lying to that officer," he said.

A couple of weeks later I was sitting in a school counselor's office talking about what I would need to score on the LSAT to get into UCLA Law when something clicked. I didn't want to study for that test. I didn't want to go to school for three more years. I wanted to chase the news. I couldn't say why exactly. I still hated the camera. I still ran away from it whenever my dad pointed it at me. But suddenly, it all seemed like a lot more fun than pushing paper behind a desk in some faceless office building.

I told my dad about my decision over lunch one afternoon in Santa Barbara. I thought he would be excited. Instead he was furious and condescending. "You might as well practice 'do you want fries with that,' " he said, "because you're never going to make it." He thought the Tur name would get me blacklisted. I thought he was treating me like a child. The fight continued all the way to my front door, which I slammed in his face. We didn't talk for a week.

I graduated in June 2005. In July, I walked into the KTLA newsroom. It smelled like must, dust, and videotape. Exactly like it did when I was little. I was home.

TRUMP
CLINTON
TOO CLOSE TO CALL

TRUMP VICTORY PARTY

NEW YORK HILTON MIDTOWN
5 P.M., Election Day

"Press or VIP?"

That's the question for Anthony and me as we enter Trump's Election Night headquarters.

"Both," I say, joking.

"Press," Anthony says, throwing me a look.

He shows the nice woman at the check-in table our credentials and smiles so widely, I think he might show her his molars, too.

"Great," she says. "Any line."

She points us to a few rows of Secret Service agents with metal detectors.

We file in behind a chatty woman in black heels (sky high) and a red dress (short). She's a VIP.

"Isn't this exciting?" she gushes. "I can't believe it's Election Day!"

It's early, but Trump's guests are already starting to show up. If the predictions hold up, it could be a short night for them, too.

Trump has brought the press to eleven different Trump properties, for more than thirty events, but for some reason tonight's party isn't inside a building with his big golden name on the front. I know that the lobby of Trump Tower was not an option—New York City already fined him ten thousand dollars for shutting the public atrium for private campaign events—but he's got a bunch of other properties nearby, including a couple of sprawling golf clubs.

So why are we at the Hilton? Perhaps because it's three blocks from Trump Tower? Or maybe because this was where Trump announced Mike Pence as his running mate, in a rambling thirty-minute speech where he talked more about himself and his personal success than he did about the man he was choosing to help run the country? Or maybe Trump couldn't decide where he wanted to be tonight and this was simply the biggest space they could find with a week's notice? I don't know.

And don't misunderstand me. The Hilton is nice. It's been host to many grand events. But it can't hold the kind of ten-thousand-person rallies that Trump has built his campaign around. And there isn't any theme or larger symbolism, as there is for Clinton's event at the Javits Center on

the west side of town—with its literal glass ceiling (get it?). There isn't even free booze. The bar is charging seven dollars for sodas, eleven dollars for beers, and thirteen dollars for mixed drinks. Trump's advisers claim that Trump is just superstitious. He doesn't want to jinx himself with a big, showy event. Cynics—or, as Trump calls them, "haters"—say he's just cheap. About that cash bar: *RedState* calls it an "abomination." *GQ* rates it pure Trump: "Let history show that up until the moment his fate became official, Donald Trump remained true to himself, a serial grifter and shameless carnival barker who let nothing come between him and the opportunity to get his grubby hands on a few more dollars." The writer adds, "I can't wait until tonight is over, and Hillary Clinton is President-elect, and an angry Donald Trump insists that Melania and Barron split their UberX home before he huffily stalks upstairs and goes to bed."

That's the other reason cynics presume Trump doesn't want to spend a lot of money: this will be a concession party. Or maybe he likes the space because it's hell on the working press. There isn't a filing station in the main ballroom, which means if you're not doing a live shot, you need to be behind a curtain in another room with folding tables.

I have twenty minutes before my next hit on MSNBC, so Anthony and I find the NBC section of the press room. Not bad, actually. There's a buffet: salad, pasta, Italian cakes, pastries. I get a cup of coffee and fill it up to the very top. I don't want to think about how many cups I have a day. I'm sure a cardiologist would be horrified. I take a good long sip and immediately regret it. Battery oil. I stuff a cookie in my

mouth to get rid of the taste, then silently thank the intern who ran out to Saks earlier to buy me a pair of emergency Spanx.

The formfitting bodysuits might as well be the official sponsor of the female press corps and perhaps a few members of the male press corps, too. The hotel mirrors of America know we all need some curve correction. But try to take yourself seriously as you open a package labeled, SHAPE MY DAY HIGH-WAISTED GIRL SHORT.

Tomorrow you will cut the sugar. You will cut the fries. You will go back to the gym. You won't be afraid of mirrors anymore.

I grab another cookie before I sit down to look at my e-mails and other messages. I have 21,376 unread e-mails. Three hundred unopened text messages. Then there are the relentless Twitter alerts that pop up and unfurl in an infinite march down my home screen. Early in the campaign I developed a much-needed strategy for staying on top of every major story. It involved turning on push notifications for a handful of my fellow political reporters, each of whom offers something specific on Twitter.

Sopan Deb of CBS News logs, fact-checks, and screengrabs the highlights of every Trump rally, interview, or appearance. Andrew Kaczynski, who was at BuzzFeed until CNN stole him, uncovers a lost Trump interview almost daily. The *Washington Post*'s David Fahrenthold is doggedly reviewing Trump's charitable giving and exposing the wide gap between what he claims he gives and what he actually gives. Maggie Haberman of the *Times* breaks a Trump story

daily, if not hourly. And the *Guardian*'s Ben Jacobs—well, he makes me laugh.

They've all been immensely helpful. But sometimes they just won't shut the hell up. Like right now. My phone is a one-dimensional firework show of pings, pops, and buzzes. I can't look away. I am so addicted to this damn thing at all times, I'm starting to look like a passed-out drunk, my head hanging limply off my neck.

I swipe through my messages. NBC just confirmed that former president George W. Bush didn't vote for Trump. It is a remarkable, if, at this point, not surprising, bit of news. The last Republican president did not vote for the Republican nominee. Let me repeat that. The last Republican president did not vote Republican at the top of the ticket. This is new.

7

"Pop the Trunk. I'm Going to Run for It."

JANUARY 25, 2016
288 Days Until Election Day

My car isn't moving. I look up ahead. The light is red.

Don't worry. Once it turns green, we'll start moving.

The light turns green. My car doesn't move.

It's fine. Just take a deep breath. Some jerk is probably just blocking the intersection. He'll move.

The dashboard clock shows 6:33 P.M.

Fuck. Am I going to miss this flight?

I'm in a cab about a mile from LaGuardia Airport. I can't quite see it yet, but I know it's there just beyond this row of houses, on the other side of the parkway. All we need to do is roll three meager blocks, turn left on the overpass, and I'm there: on my way to Iowa, the first state to cast a ballot for president.

For seven months, everybody in the world of politics has

been talking about who will be the next president and why. But all of it is pure theory until Iowa.

"Sir, any idea what's going on?" I ask the cabdriver.

He shrugs. I close my eyes and try to breathe.

Katy, it's fine. We'll start moving any second.

When I open them, I see cars are peeling off ahead, abandoning their place in line. Not a good sign.

The dashboard clock shows 6:40 P.M. I check my phone: yup, 6:40 P.M.

My flight leaves at 7:27. Forty-seven minutes from now.

My winter coat suddenly feels tighter. My face is hot.

"Sir, can you turn the heat off?"

I roll down the window and begin to sway back and forth in my seat. I crane my neck to look ahead.

"Sir, turn on 1010 WINS. Maybe we can catch the traffic."

Too late. They're on to weather. The top headline is still snow removal.

Shit, shit, shit.

"Sir, is there anything you can do? Can we take another street?"

No, he says. This is the only street to the overpass.

I start to spin.

Why did he get off the parkway?! We were moving just fine. Why can't anyone do their job? No, it's my fault. I should've told him to stay on the highway. Why didn't you tell him to stay on the Grand Central?

My worst qualities are bubbling over. My mind is like a washing machine with too much detergent, spinning way too fast. I blew up my life for this assignment, believing, in the way of Noël Coward, that work could be more fun than fun. But the gamble hasn't yet paid off, and it never will if I don't make this flight. It's a Monday night, the day before the seven-day countdown to the Iowa Caucus begins. All the other reporters who are dying to take over the Trump beat are already in Iowa. And, given an opening, they will gladly take it.

Fuck. Fuck. Fuck. Fuck. Fuck.

"Pop the trunk," I say. "I'm going to run for it."

My suitcase is the size of a refrigerator, and it is fighting me down this godforsaken side street in Queens. The wheels weren't made for snow and it's too goddamn big. I didn't want to take it. I have a rule: carry-ons only. I don't have the time to check a bag, because I don't have the time to watch a banged-up black belt spin in a circle from now until eternity. I'm not Sisyphus and this isn't purgatory.

Or is it?

Damn this damn bag. I hate it. I'd like to leave it on the side of the road like a rotting couch. But there's no telling when I'll get back home to repack. Not "home." Forget home. For six months I'd been living at the Standard hotel, the quieter one on the Lower East Side. Sure, it sounds luxurious. Room service, maid service, the balm of a perpetually stocked minibar. However, it turns out there's only so many bags of Smart Puffs you can wash down with a shot of Hendrick's before you start to long for a proper kitchen. Jeez, by the end I would've settled for a hot plate. Instead, just a few weeks ago, I convinced a friend to let me stay at his place. Truth be told, it was not that much of an upgrade. The room at the Standard was 250 square feet. His apartment is 400 square feet—half of which are now covered with my clothes.

I'm breaking my carry-on rule because I have to pack for at least a month, maybe longer. I also have to pack for multiple climates. On February 1, the Iowa Caucus will be cold. On February 9, the New Hampshire primary will also be cold. But on February 20, I'll be in South Carolina, where it is warm. And on February 23, I'll be in Nevada, where it is even warmer. You get the picture. If Trump wins or, at the very least, doesn't drop out, I'm going to need every scrap of clothing in this bag.

Did I just drop something?

I turn around. It's dark and there is snow everywhere. Lots of it. Two days ago, the city got more than two feet, and most of it is still on the sidewalks. If I did drop something, and it's likely I did, there's no way I could find it.

Please don't be my portable Wi-Fi.

I've already lost one of those.

I dig my phone out of my jacket, a giant red parka.

Thirty-two minutes. Pick up the pace, Katy.

My bag is not a bag anymore, it's a child throwing a tantrum. Every time I spin around to give it a good yank I can feel my backpack opening. I've jammed it with notebooks and peanut butter packets, all of which may be dropping out like seedlings I'll never see again.

Is that a shoe? The toe is facing up. Jesus, did someone get buried?

The overpass that leads to the airport is crowded. Fellow travelers just as desperate as I am are pushing and dragging their belongings in the snow. With all the foot traffic, it's really just slush now. Big dirty heaps of it. Someone once said that New York is the only city that makes its own gravy. And this is New York gravy at its finest. Cold, brown, chunky, gross.

"Ahhh!"

I catch myself before my face lands in it. My legs aren't as lucky. I've fallen for New York City's favorite prank: the camouflaged curb. I'm soaked up to the knee and my shin is throbbing. I hit something down there. Don't ask me what. The curb? An open sewer grate? A discarded toaster? I look down. My pants are ripped and I'm bleeding. This is either going to kill me or spur some sort of superhuman slush-based mutation. (Fingers crossed, it's the ability to fly.)

No one tries to help me. In fact, people are looking at me as if I'm crazy. Probably because I look crazy. I'm dripping wet from the waist down. I'm talking to myself. Charging cables are hanging out of my backpack like a makeshift bomb. And I'm walking into an airport with a suitcase the size of a Ford Bronco.

Trump calls LaGuardia Airport a third-world country. How does he even know? For him, there are no terminals, no gate agents, no boarding groups, not even hard departure times. The extent of Trump's complaints about LaGuardia are what he can see out of his private jet window. He gets to drive right onto the runway, right to the stairs of his very own 757 with the twin Rolls-Royce engines. He doesn't have a coach seat. He has a white leather captain's chair with his initials embroidered in gold.

My own seat is somewhere at least two terminals away. All around me people are blowing their stacks: stranded, delayed, freaking out into their phones and the thirty-degree air. It's like I'm back in Jakarta, except for the weather and the font on the license plates of all the cars going nowhere.

I run through a lattice of sidewalks best described as postearthquake and finally arrive at the Delta check-in. There is no line, because evidently no one can get to the airport today. The agent looks up as I arrive, panting and red-faced.

"I'm on the flight to Des Moines."

"Tonight's flight?"

"Yes."

She doesn't look at her screen. She just knows.

"That flight takes off in twelve minutes."

"Yeah, I know. I have my ticket on my phone . . . but I also have this bag I need to check."

We pause and gaze at the Montana-sized suitcase at my feet.

"Again, ma'am, that flight leaves in twelve minutes. We stopped checking bags."

I try to reason with her, never sounding less reasonable myself.

But I'm Diamond . . . but you can't take off if no one can make the flight . . . but you can send my bag on the first flight tomorrow . . . but I have to be on TV in the morning!

The agent confers with a manager for a minute or two, then hands me a paper ticket and says, "If you can get that bag through security, you can try to make the flight. But the gate is already closed and you're not making the flight."

I thank her and start to run, newly energized. At security, there is no line, because, again, no one can get to the airport. That's the good news. The bad news is the TSA agents get a good, long look at me and my bag. They're puzzled, then amused, then outright rooting for me. I'm not sure I can lift the bag onto the belt, let alone get it into the X-ray machine. But I try.

It scrapes through and I practically leap through the metal detector to meet it on the other side. I'm not sure whether the agent is studying the bag on the monitor or if the bag is stuck. It's six minutes to takeoff when the bag is back in my hands, now deemed safe for flight.

Holy crap. I can't believe they let me in with this. I'm going to make this flight. I am going to make this flight.

My internal monologue sounds like the Little Engine That Could—*I know I can, I know I can*—as I run to my gate. I get there with three minutes to spare. The plane is still there, but the gate area is empty. The boarding door is closed and the counter is deserted. It's just me and another straggler, looking at each other, toddlerlike, searching for an answer in the other's eyes. We need to find someone. Anyone.

"Do you see anybody?"

The guy looks at me helplessly.

Useless.

I look around. The only other person in sight is a guy cleaning up.

"Can you call down to the gate and see if they can let us on?" I ask.

"What?" he says.

"Can you call down to the gate and see if they can let us on!" This time it is more a demand. Either he empathizes with my desperation or he is worried about provoking a crazy woman. Either way, he calls down.

"Someone is coming," he says, hanging up the phone.

Two excruciating minutes later a flight attendant appears from behind the closed door.

"Can you please, please, please let us on," I ask, pointing at the other traveler behind me. The useless straggler.

The agent looks at my bag—my motherfucking shipping container of a bag—and down at the flight's passenger manifest and laughs.

"You must be Katharine Tur. Was wondering if you'd make it. Sure. Come down. I'll stuff that in the flight attendants' closet," he says, pointing at the bag.

I take a deep breath and thank him enough times to make it awkward while I board.

Safely in my seat, I dig my phone back out of my coat and write this down. No one is going to believe me.

It's 7:27 P.M.

A week later, Trump loses Iowa.

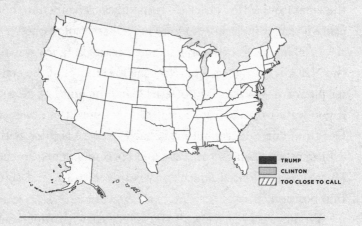

TRUMP
CLINTON
TOO CLOSE TO CALL

TRUMP VICTORY PARTY

NEW YORK HILTON MIDTOWN
5:45 P.M., Election Day

Behind the barriers, please, behind the barriers," a Trump volunteer is saying. "The media has to stay behind the barriers."

I'm at the gate to the floor of Trump's election headquarters, hoping to snap a picture of the press riser, which is so well lit that it looks like a portal to the fourth dimension. All the major news outlets are here, along with dozens of local stations, and international teams from places like Japan, France, Germany, Britain, and Israel. The whole world is watching—just not from the vantage point on the ballroom floor. The campaign doesn't want us mixing with the voters and VIPs. Or else Trump likes us penned in, because it makes it easier to point and yell at us all at once.

I roll my eyes at the woman and start making my way to the NBC live position. We're on the lower level of the center platform—to the right of Carl Cameron and Fox News, if you want to make a joke of it. I'm no more than twenty feet from our camera, but every inch of space is filled with cables, other people's cameras, and correspondents, so the journey is perilous. I squeeze, twist, duck, turn, and spin into position.

We have a great view of the stage, set with American and state flags. Off to each side, there are two glass cases, which for some reason contain museum-quality examples of Trump's iconic campaign hats, one in white and one in red. The ballroom floor itself is divided into two sections, one for general admission guests, the other—dotted with cocktail tables—for VIPs.

I plug in for *Meet the Press Daily* with Chuck Todd. Across town at the Javits Center, Kristen Welker is doing the same. The plan is for us to deliver dueling reports about what each campaign believes is their winning path. Clinton has a bunch of them, but Trump has to win Florida. There is no way around it. Now, if you ask Trump staffers, they publicly say the campaign is confident and so is Mr. Trump.

But that's not the way Trump sounds to me. Besides musing about a long vacation, he keeps claiming Clinton and her supporters will cheat.

Watch your polling station, he warned. Democrats will vote twice. There's dead people on the voting rolls. The election is rigged!

Now he thinks he has some evidence. "Utah officials report voting machine problems across entire country," he tweeted at 4:28 P.M., incorrectly citing a CNN report that mentioned a single county, not the whole country. Sure, it might have been an innocent error, but Trump has also laid the groundwork to cry "conspiracy" if he loses. There's "a global power structure," he says, designed to rip off the working class with Clinton at the helm. She's in "secret" meetings, he says, "with international banks to plot the destruction of U.S. sovereignty in order to enrich these global financial powers, her special interest friends and her donors."

I run through Trump's path to victory and toss back to Chuck. Back in the press room, Ali Vitali says the early returns don't look good in Florida. Her source says it's the RNC's fault. The party didn't do enough to help Trump, didn't put enough into the get-out-the-vote effort. A minute later, Kellyanne Conway tells Chuck she's "disappointed" she didn't have the full support of the Republican Party.

Here come the knives!

I tweet: "Publicly and privately we are already hearing frustration from Trump camp about the coordination bw RNC and Team Trump. @Alivitali."

My phone rings while still in my hands.

Never good.

Hey, I say, trying to sound at ease.

Casual Katy.

It's Sean Spicer, the communications director for the RNC.

Who told you there was frustration between the party

and the campaign? he asks, in a tone somewhere between a growl and a yell. It's not true.

I start to tell him about Ali Vitali's reporting, but he interrupts.

Katy, I'm telling you it's not true.

Sean, it's not just Ali's source.

Who's saying it? he demands.

Well, this is easy.

Kellyanne Conway just said it on the air with Chuck Todd, I say.

What? Huh? What did she say?

Read my tweet, I say. She said she was disappointed she didn't get the full support of the party.

Sean is silent for half a second.

I gotta go, he says.

Click.

8

"Look at Those Hands.
Are They Small Hands?"

MARCH 3, 2016
250 Days Until Election Day

It's twelve days before the Florida and Ohio primaries. The smell of beer and fried food is teasing the hundreds of journalists inside the Hockeytown Cafe, a sports bar in downtown Detroit. Tonight there is no beer. The bar is closed. There are no sports, either. The walls are decorated with hometown memorabilia, but the big screens are all tuned to Fox News, which is minutes away from starting the eleventh—repeat, eleventh—Republican presidential debate. The first three debates were arguably a valuable window on what the candidates believe and how they think under pressure. It's been a pulp show since then.

The field has been winnowed from sixteen candidates to

four, Marco Rubio, John Kasich, Ted Cruz, and Trump, but they've spent the extra time throwing punches more than parsing policy. Why? In a word, Trump. You can't dance the flamenco with a sumo wrestler. Trump comes into tonight with ten primary wins, including seven states on March 1, aka Super Tuesday. If Trump's rivals are going to stop him, they have to do it in these next twelve days.

Tonight is their first big chance.

Moderators Megyn Kelly, Bret Baier, and Chris Wallace introduce the candidates and the debate begins. I'm on the third floor of Hockeytown, surrounded by tables of NBC News journalists. The place is big, and somewhere below me in the glow of laptops and TV screens are just about all the nation's traveling political journalists. Across the street is the grand old Fox Theatre, where the debate is happening. The theater is open only to Fox News reporters. So we're in the "press file," a make-do place for the rest of the working media. A generation ago, hell, even four years ago, that meant a place for print reporters to file text pieces while TV reporters script for packages or pop outside for stand-ups and live shots. Now, we're all feeding the same multiheaded dog of digital: Web sites, social media, whatever's next.

To turn a drinking place into a working place, the Republican National Committee and Fox News have made some small adjustments. They boosted the wireless Internet, added 250 electrical power strips, and, perhaps less helpfully, turned off the beer taps. They also laid out a buffet, described as roasted corn salsa and chicken salad; boneless chicken with caper and artichokes with white wine sauce; braised beef short rib with red wine demi; crispy panfried Kalettes with sea salt and Parmesan; ziti with black pepper, butter, and Romano; and whitefish with triple lemon butter.

The menu is mouthwatering. The food is not. Still, I'm going for seconds and lightly cursing the line, filled with shuffling journalists, each a bit rounder than they were ten debates ago. All this food and extra technological oomph comes in handy right off the top, when Chris Wallace brings up the big news of the day: Mitt Romney's speech at the University of Utah.

The 2012 Republican presidential candidate took the 2016 Republican front-runner apart with a hatchet. Romney called Trump a "con man, a fake." With a Trumpian flourish he also mocked the candidate's foreign policy as "very, very not smart." Then he brought up Trump's many, many failed businesses, too, including Trump airlines, *Trump* magazine, Trump Vodka, Trump Steaks, Trump Mortgage, and Trump University. He was not endorsing another candidate. Rather, he was begging voters to choose one of the three *other* GOP contenders. It didn't matter which one, just so long as it wasn't Trump.

This went on for seventeen minutes. It was brutal.

"Mr. Trump, as you may have heard, the 2012 Republican nominee for president, Mitt Romney, had some things to say about you today," Wallace says.

The crowd cheers and boos and carries on. The crowds at debates aren't supposed to make any noise at all, but they never listen to that rule. People are people.

"He challenged you to answer with substance, not insults. How do you answer Mitt Romney, sir?"

I consider for a moment that Trump might actually take the higher ground.

"Well, look, he was a failed candidate," Trump says.

Nope.

"He should have beaten President Obama very easy. He failed miserably and it was an embarrassment to everybody, including the Republican Party."

Hockeytown reacts like a rowdy movie theater crowd. The plot is just too predictable. Trump says he's a counterpuncher. But it goes deeper than that. He is the Pavlov's dog of politics: insult him and he'll insult you back. He has to respond even if his "counterpunch" lands in his own face.

Trump finishes his answer, which is more of the same, and I start thinking about the spin room, a place where candidates and campaigns go postdebate to talk up their performance. This Romney spat feels like Topic A, and I need to get Trump to respond. That's a big part of the job on debate nights. I jot some potential questions, send a few tweets.

Now Bret Baier is bringing up Rubio's sudden turn toward Trump-style insult politics.

"In the past week you've mocked Mr. Trump's tan. You've made fun of his spelling. You called him a con artist. You suggested he wet himself backstage at the last debate, along with other vulgar jokes and jabs," Baier says, riding another wave of crowd reaction. "So what happened?"

Good question. Trump's rivals have been trying to differentiate themselves as the adults in the room, the only ones you can trust with the nuclear codes. But Baier's question is incomplete. You see, a few days ago, Rubio was at a college in southwest Virginia, where he got himself on a real comedic roll. He brought up the subjects Baier mentioned, but he also alluded to one more: Trump's manhood.

"He's like six foot two, which is why I don't understand why his hands are the size of someone who is five foot two," Rubio said. "And you know what they say about men with small hands . . ."

He answered his question with a folksy "you can't trust them"—but the crowd knew what he meant.

So did Trump.

So did the rest of us in Hockeytown.

Onstage now at the Fox Theatre, Rubio doesn't mention his manhood comments. Instead he says, in essence, that Trump made him do it. The mogul deserved the insults, Rubio argues, given the insults Trump has thrown at his opponents. "Let's have a policy debate," he says.

"And we will," Trump jumps in.

After a few more pleas for decency and intellect, Rubio rests.

"Mr. Trump, your response?"

Hockeytown takes another bite of food, another sip of coffee.

"Well, I also happened to call him a lightweight, okay? And I have said that. So I would like to take that back. He is really not that much of a lightweight. And as far as . . ."

Trump stops midsentence, as though remembering something Bret Baier forgot.

". . . and I have to say this, I have to say this. He hit my hands. Nobody has ever hit my hands. I have never heard of this. Look at those hands. Are they small hands?"

Hockeytown pauses midchew, midsip.

"And he referred to my hands, if they are small, something else must be small. I guarantee you there is no problem. I guarantee."

Hockeytown chokes and splutters.

Did he just say that? Did he just say THAT? DID DONALD TRUMP JUST DEFEND THE SIZE OF HIS PENIS IN A PRESIDENTIAL DEBATE?

"Okay," Baier says. "Moving on."

Hockeytown cannot move on.

There's another ninety or so minutes of scheduled debate time, but short of an onstage heart attack, penis size is now the story. I toss my Romney-related notes to the side and try to think of a way to ask Donald Trump about, well, I shudder even to write it down. Can I say *penis* on TV? What about *manhood*? No, no, no. Trump may be willing to test the voters' patience with his lack of decorum, but I am not willing to do the same with viewers.

With twenty-five minutes to go, Hockeytown starts to stir. The nation's political journalists close their laptops, pocket their phones, and climb back into their winter parkas. Usually the spin room is either in or attached to the "press file." But there's no getting Trump, Rubio, Cruz, and Kasich across the street to Hockeytown. Instead we have to face the literal blizzard that is raging outside to confront them on the metaphorical blizzard that Trump's penis reference has created. We scurry into the lobby of the Fox Theatre.

It's beautiful and gigantic: thirty-five hundred square feet of floor space, six stories tall, all of it dripping with Egyptian, Indian, and Oriental designs. The marble floors are punctuated by marble-esque pillars and the ceiling is sculpted in gold—so much gold I can't help but wonder if Trump will be jealous when he comes out. Right now he's somewhere deeper inside the theater, but at any moment he is expected to appear at the top of a twin staircase at the end of the lobby.

As soon as I see the setup, I start casing it. I need to predict where Trump will go once he gets here, because I need this interview. I need it for tonight's coverage. I need it for tomorrow's coverage. I need it for a *Nightly News* package. I need it because I don't want to get scooped by a competitor and I don't want to get scooped by one of my colleagues. I need it because if Trump keeps going in this race, I intend to be the lead reporter on his tail. I've worked too hard not to be. I've given up too much not to be. No, Donald Trump is not getting out of this room without taking my questions.

At the foot of the stairs, below twin statues of some mythical lion creature, is an elegant red velvet rope intended to keep Hockeytown's rabble a safe distance from the candidates. The rope stretches from the front of the stairs all the way down along the right side of the lobby. So that's my choice: I can be on either end, not both. Most of the press is clumped right in the middle. So I gamble he'll hang a right off the stairs, so he can start from the beginning, maximizing his camera time by walking the full length of the rope.

I'm right. *Thank goodness I'm right.* Here he comes with Melania, walking right to my position. Now I just need to get his attention. My cameraman is tall. That helps. But after eleven debates and countless interactions, I've learned that the key to grabbing Trump is to yell out a compliment—usually something about a positive poll.

But tonight I don't even have to yell. Trump sees me and, evidently in a good mood, he walks right over. Now the challenge is keeping him. I'm the first person he talks to, and since I want to keep him talking for as long as I can, I start with a softball.

"How'd you feel about tonight's debate?"

"I thought it was great. I really had a good time," he says. "I did think it would be more evil. I did think it was going to be more vicious."

More evil? I still don't know how I'm going to bring up his, um, hands. So I just keep the conversation going.

"What about the schoolyard taunts of the debate—"

"I don't see schoolyard," Trump interjects.

"There were references to hands?" I say.

There, I've done it.

"Yeah, but see how beautiful my hands are. Look at those hands, those are powerful hands."

He shows me his hands. Melania smiles behind him. Trump turns to how useful his hands are in action.

"Hit a golf ball 285 yards. I was very proud to hold my hands up, because these are politicians. They say things and they make things up, and you know you have to clear up the record."

March 8. Seven days before the Florida and Ohio primaries, seven days for Trump's rivals to sideline his rise. It's midafternoon in Jupiter, Florida, and I'm in the back seat of a rental car, trying to change my clothes. I'm ditching my jeans and T-shirt for the skirt and blouse I didn't have the energy to fuss over when I got out of bed at five this morning. I had to catch the first flight out from Mississippi to Florida, where Trump is holding a Super Tuesday Round 2 party at his National Golf Club in Jupiter.

Four states are voting in the GOP primaries tonight: Mississippi, Michigan, Idaho, and Hawaii. Forty-six other states will watch and wonder if Trump fever has started to break. Over the weekend Trump lost Maine and Kansas to Ted Cruz. This morning two new polls showed Cruz gaining ground. The professional prognosticators have rubbed those two sticks together to make a fire.

After I change, I check in at my live location, on the courtyard between the eighteenth hole and the club's ballroom. Then I have to see the ballroom where Trump will hold his victory party. And, my word, it's spectacular. I don't mean the architecture or the design. I mean the food: a table of fancy finger food, including the campaign trail's rarest food form, fresh vegetables. Carrots, celery, peppers, and tomatoes, just washed, glistening in the twilight of a Florida sunset. It's so beautiful I want to cry.

I wonder if Trump's staff made a mistake by putting this out where the journalists have access to it. There's an open bar, too, all the liquor and beer you can drink. It's nice not to be penned in for a change, but this is weird. The ballroom is filled with rows of chairs, the kind you'd sit in at a fancy hotel wedding, wood painted gold. The first few rows are reserved for guests, leaving Trump a healthy buffer between himself and the press. He doesn't want to get near us, but he hopes we'll eat our veggies?

Never mind. Something far weirder is next to the stage. It looks like piles of raw steaks, bottles of wine, pallets of water, and a propped-up magazine. The Secret Service is guarding it like a table of ancient artifacts, which in a sense it is. The wine is from Trump Winery in Virginia. The water is wrapped in baby-blue Trump logos. The magazine, called *The Jewel of Palm Beach*, has a Trump property on the cover. And the steaks? They appear to be Trump Steaks.

This is bizarre even by Trump standards. So I send a note back to the entire politics team at NBC, and by the time I look up from my phone a crowd has gathered. The whole press corps seems to be poring over this QVC-esque display. And we start to notice some issues with it.

"They're not even Trump Steaks," one reporter says. Sure enough, the plastic wrap is emblazoned with BUSH BROTHERS, the name of a butcher in Palm Beach. The real Trump Steaks, once available via Sharper Image, aren't for sale anymore.

"Does Trump own a bottling plant?" another reporter asks.

No, he does not.

The water is from Village Springs in Connecticut. Again, this is not deep investigative work. It says VILLAGE SPRINGS in small print on the label. And Village Springs is a company that sells water with any label you want to put on it. That's the business. Trump simply paid them for the labels.

Same with the magazine. It's a glossy brochure published by a company called the Palm Beach Media Group. Like the water bottler, it is another company that offers customizable content. You pay, they print. That's what Trump did.

The wine bottles are the most unintentionally funny. They are emblazoned with Trump's name. But for some reason—tax benefits?—a disclaimer on the winery's Web site says, "Trump Winery is a registered trade name of Eric Trump Wine Manufacturing LLC, which is not owned, managed or affiliated with Donald J. Trump, The Trump Organization or any of their affiliates."

So there you have it. This display is clearly Trump's counterargument to Mitt Romney, who itemized the mogul's many failed businesses. But the display actually proves Romney's point. None of the items on the table, not one, actually represents an ongoing business of Donald Trump.

Two or three hours later, the polls close and Trump is a big winner again. He takes Michigan and Mississippi, and things look good for him in Hawaii. He takes the stage just after 9 P.M. and congratulates himself. Then he launches into what has clearly been irking him since last week.

"Mitt Romney got up and made a speech the other day,"

he says. The Trump supporters in the room offer a round of supportive boos.

"No, it's okay," Trump says, mock-soothing them. "He's a very nice man." The supporters laugh. And then Trump looks over to his props.

"I brought some things."

Oh, here we go.

"He said water company is gone. I said, it is? I didn't know that. I have very successful companies. Let me just explain."

Yup, he's doing it.

"Trump Steaks—where are the steaks? Do we have steaks?" Trump says, looking around for the red meat. "We have Trump Steaks. And by the way, if you want to take one, we'll charge you about, what, fifty bucks a steak?" he jokes.

I look over at Anthony. He is laughing so hard I have to nudge him hard with my elbow. "It's a Trump infomercial," he squeals. His eyes are watering. I don't think he can breathe.

"We have *Trump* magazine. Let me see the magazine," Trump continues, oblivious to or, more likely, unconcerned by the fact-checks that are lighting up social media.

"It's called *The Jewel of Palm Beach* and we—it goes to all of my clubs. I've had it for many years and it's the magazine. It's great. Anybody want one?" He throws one to a supporter up front.

There's no airplane next to him, but Trump launches into a defense of his defunct shuttle anyway.

"I sold the airline. . . . The economy was horrible and I made a phenomenal deal."

Now Trump University—which is facing a fraud lawsuit.

"So we're putting it on hold. If I become president, that means Ivanka, Don, Eric, and my family will start it up. But we have a lot of great people who want to get back into Trump University."

He almost forgets the wine.

"And by the way, the winery . . . it's the largest winery on the East Coast [it isn't]. I own it 100 percent, no mortgage, no debt [he doesn't]. You can all check [we did] . . .

". . . And I know the press is extremely honest, so I won't offer them any, but if they want they can take a bottle of wine home."

Anthony and I exchange a knowing look. We could use a drink. But a little while later, by the time I look back for that bottle, the bottles are gone. Trump's staff has whisked them away.

March 15. The two weeks are up. If Ted Cruz, Marco Rubio, or John Kasich has any hope of stopping Trump from getting the nomination, one of them has to win Florida or Ohio tonight. Those are winner-take-all states with a combined 165 delegates. Governor Kasich is popular in Ohio, his home state, but he hasn't won a single primary yet. Rubio represents Florida in Congress, but Trump is beating him in the polls.

So it's not looking good for anyone but Trump. For tonight's watch party, he's picked his Mar-a-Lago resort in Palm Beach. It's a mansion he hopes to one day call the Winter White House, which, in fact, is what the original owner—cereal heiress Marjorie Merriweather Post—hoped it would be called, too.

In her will she donated the 128-room, seventeen-acre estate to the federal government, hoping to entice future presidents to winter there. The Nixon, Ford, and Carter administrations didn't use it, though, and in 1981, under President Reagan, the government gave the property back to the Post Foundation, citing the million-dollar annual maintenance fees.

Trump bought it four years later, for five million dollars, two million less than it cost Post to build the place in 1928. But by the mid-1990s, it was Trump who couldn't handle the maintenance fees. So he turned the estate into a private club, kitted out with a twenty-thousand-square-foot ballroom covered in what seems like all the gold he could afford.

That's where I'm standing. We'd come to Mar-a-Lago before, for a news conference. But this is Trump's first campaign use of the full ballroom. It's a lot like the last Super Tuesday event—but a lot has changed as well. The freedom we had to inspect Trump's steaks—that's gone. We're roped off in the back of the room. No fancy finger foods. No open bar. No refreshments at all except for a selection of sodas, room temperature.

It's also much fancier than Jupiter. There are rows and

rows of chandeliers. The stage has shimmering curtains, glowing with red, white, and blue light. And the people, my goodness, the people. It's as if Trump's most glamorous friends and club members have all shown up for a gala at the Metropolitan Museum of Art.

It's actually kind of impressive. Trump launched his campaign with paid actors in T-shirts. Now he has real people in silk ball gowns and men in six-thousand-dollar tuxedos lightly pushing and shoving to get face time—some having paid a one-hundred-thousand-dollar membership fee for the right to be here. It's an amazing reversal. Trump has gone from paying Joe Schmos to cheer for him to accepting money from Joe Somebodies who will happily do the same.

But as I watch all this money walk around, as I survey a room of people nipped, tucked, and sucked to their ideal of perfection, I can't stop thinking of Trump's rally crowds. The people in this room are decidedly not the people at his rallies. The rally people arrive in denim, flannel, and thick-soled boots. They wait for hours, eat whole pizzas in the security line, tattoo Trump's face on their forearms.

The people in this ballroom are not the subject of Trump's speeches, either. Their industries aren't dead. Their jobs didn't disappear overseas. More likely, these are the people shipping the jobs overseas. These are the people slashing budgets and enhancing their own bottom line while the bottom falls out of everyone else's lives.

What would the people at Trump's rallies say about the people at his victory parties? What would the folks who are fanning Trump's political flames think of all these gilded types trying to warm themselves by Trump's new fire? I put the question to a man in a tux, who identifies himself as Trump's friend. "Why do people fighting for a raise relate to all this?" I ask. "Because deep down," he says, "they know he's one of them."

Do they? Then what do all these rich people think about Trump?

I think back to a conversation I had with a billionaire, someone whose wealth comes from business, not the performance of business. Trump is "a dangerous joke," he said. You see, there are not that many billionaires in this country. Most know each other or know of each other, and Trump, with his reality show flair, was a bit of an oddity. In the eyes of some other billionaires, he wasn't a real billionaire at all.

But now, well, that's starting to change. As he gets closer to power, money is getting closer to him.

I'm distracted by a woman with a sapphire the size of a dinner plate hanging from a rope of diamonds around her neck.

"That's quite a necklace," I say, and she shakes her head at the comment. The sapphire, she says, ruefully, isn't that good. "It should be darker."

What do you say to that? I have no fucking clue.

But maybe the tuxedoed man, my old billionaire contact, and the sapphire lady are all telling the truth. Maybe the necklace really is a piece of junk, Trump really isn't a billionaire, and, most important, his supporters really do relate to him. He's rich, but he's relatable rich.

Trump appears onstage with his campaign staff, the Bad News Bears of Politics, the people the other candidates didn't pick. Last week, one of them—Trump's campaign manager, Corey Lewandowski—was accused of assaulting a female reporter. It allegedly happened right after Trump's press conference with the steaks and wine, as Trump walked through a room of reporters. I was on Trump's left when something happened across the aisle—on Trump's right. Breitbart reporter Michelle Fields said she tried to ask him a question, but couldn't—because she was violently jerked backward.

She claimed Lewandowski had done the jerking, pulling her hard enough to bruise her arm. She posted a picture of the bruises on social media. But Lewandowski denied the incident and the Trump campaign went on the offensive. They implied she was a liar and opportunist, "an attention seeker" who's done this before.

The incident is a big political talking point because Trump has a problem with female voters. Most don't like him, according to polls. He has a long history of saying ugly, objectifying things about women, which his opponents are using against him. So when Lewandowski was accused, most observers seemed to think Trump should tread carefully and perhaps err on the side of believing the woman involved. A *Washington Post* reporter had already come forward to say that he had witnessed the incident and that Fields's claim was accurate.

But Trump opted for blind defense of a man.

I asked him about this a few nights ago in the spin room

after another GOP debate, this time in Miami. He ignored me at the front of the rope line, so I went to the end of the rope line and plotted to ambush him before he walked out. It was nerve-racking, waiting there, in position but out of sight. Trump was on the line with Lewandowski himself, who had already seemed to be on the verge of freezing me out entirely. But this was too important to ignore. I had to get Trump on the record.

Just before Trump reached the doors, I popped out. With Lewandowski in the frame of my camera, I tossed a couple of softballs first. Then, as Trump's eye drifted toward the exit, I went for it.

"On the story of your campaign manager, Corey, and Michelle Fields, I know you just said . . . she may have made the story up. The *Washington Post* reporter is the one that reported it and saw it; do you think he's making it up as well?"

Trump claimed not to know "anything" about the incident, but then he proceeded to defend the accused.

"I can tell you, Corey's a—he's just an incredible guy, I don't think he would ever do a thing like that."

"She posted a photo of bruises on her arm," I said.

"Well, I don't know, how did they get there, huh? Do you know how they got there?"

"I'm just saying what's being reported and what's being alleged right now."

"You tell me how they got there. I have Secret Service agents all over, I have cameras all over, nobody saw it happen, nobody complained, and I know this guy, and he's a fantastic guy."

"If it comes to light that something did happen, are you going to do something about it? Will there be some sort of punishment?" I asked.

Trump's face hardened at this question.

"Supposing it comes to light that nothing happened, are you gonna apologize to him for what you're doing?" he said.

"I'm not reporting anything that I saw personally, I'm reporting what Michelle Fields is reporting."

Trump turned to his campaign manager and back to me.

"Will you apologize to him if it turns out that nothing happened?" he asked, again.

I started to say that I'd report what happened, certainly, but at that, Trump walked away and Lewandowski followed.

Now in the ballroom five days later, with Trump awash in another big Tuesday night, Lewandowski is right there next to him. Ohio went to John Kasich, but Trump won Florida, Illinois, and North Carolina, collecting a huge number of delegates. "Good job, Corey," Trump says.

This is not a candidate distancing himself from a troubled aide. This is a man pulling that aide closer. Trump won big on March 8, despite a broadside by the last Republican presidential nominee and despite a debate that featured him volunteering a defense of his private parts. He won big tonight, despite bald prime-time lies about the supposed success and continued existence of Trump Steaks, water, and wine. Now, with this "fantastic guy" by his side, he's betting he can keep on winning.

Two weeks later Lewandowski is arrested and charged with simple battery. A surveillance video confirmed Michelle

Fields's version of events—Lewandowski grabbing her, pull-ing her away. Does Lewandowski apologize? He does not. Does Trump punish him? He does not. Does Trump keep on winning? He does.

April 6. I'm back in London trying not to cry. I was last here at the end of 2015, when I got a call from the president of NBC News. She said, "We want you to move back to New York."

I said, "For what?"

She said, "To continue doing what you are doing?"

I said, "Does that mean I'm the lead reporter on Trump? Does that mean I'll be covering politics even if Trump loses? What do I do when this is over? What is your plan for me afterward?"

She said, "We haven't gotten that far."

To cover the Trump campaign, starting last June, I sac-rificed a vacation, a boyfriend, a dream job overseas, and in many ways, stability. But these are the normal fatalities of ambition. Lots of people torch their personal lives for a shot at professional success. But now, moving box by moving box, I am losing something else before my eyes. Not just a personal life—but a life itself, a way of being in the world.

To make matters worse, my movers keep pausing to talk to me about the person they see as the ugliest of ugly Americans, the one somehow making a successful bid for the presidency. They can't believe it's happening.

"What 'beans' is he going to spill about that other guy's

wife?" they wonder, referring to a bizarre Trump tweet about Ted Cruz's wife, Heidi.

"Why do Americans like him?" they ask.

"Does he really have a chance?" they want to know.

I don't know, I say. Because they're frustrated, I offer. Yes, I respond.

I'm not allowed to touch the boxes for customs reasons, so I concentrate on sorting trash from treasure in the first flat that I ever felt was all my own. They say London is gloomy and overcast, but in my bedroom, every morning the sun would somehow find an angle into the room and linger there.

Now it's gone, and as the hours roll by, the flat empties out, until about 3 P.M.—when there are no more boxes, no more Trump questions, and no more *my flat*. On the street, I look back toward the windows, where my lavender is blooming in the box. It still feels like a night out, not a last exit. But at my friend's house, two seconds after walking in the door, I break down at the realization that I am no longer a Londoner. I'm an American in a friend's guest room.

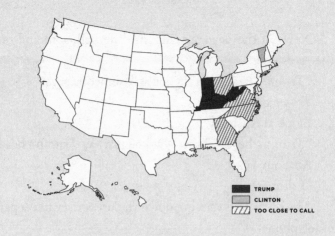

TRUMP · CLINTON · TOO CLOSE TO CALL

TRUMP VICTORY PARTY

NEW YORK HILTON MIDTOWN
7:30 P.M., Election Day

Donald Trump's head is rolling by on a platter. The polls are starting to close and he's just won a handful of red state giveaways: Indiana, Kentucky, and West Virginia. Now his head is arriving from Trump Tower "to celebrate." Or so says the man pushing the cart. People are staring. Confused. Trump's head doesn't look happy. He's just won three states! But instead of a smile, there's a scowl. And there's something weird about his eyes. They're glazed over. Lifeless. It's hard to tell if you should give him your congratulations or condolences.

News of the head spreads into the ballroom and onto the press riser.

"Trump's head is in the lobby!" someone says.

Trump's what?

I look at Anthony.

"Trump's head?" I ask. "Did he just say Trump's head?"

Anthony laughs.

"I think so."

The bulletin is moving down the press riser like a game of telephone.

"Trump's what?"

"His head!"

"Huh?"

"Whose head?"

"Trump's."

"His head?"

"There's a head?"

"What head?"

Flashbulbs are popping. A crowd is forming by the lobby doors.

"I'm going to take a look."

I zigzag out the ballroom and into the lobby.

Sure enough, right next to the cash bar, there is Trump's head. It is unmistakable. The blond hair. The duck lips.

"It's a cake!" someone yells.

"Can we eat it?" asks someone else.

"Is it chocolate or vanilla?" another wonders.

"Move over, I want a picture," says a fourth.

A line forms. Trump supporters are taking selfies. Reporters are posting pictures to social media.

You can tell who's covered Trump and who is new to this kind of spectacle. The newbies want in on the Trump-world retweets. It's a social media club.

To those of us who've been around this block, it's just another "of course." And a chance for a rally joke, a winking reference to Trump's justification for his antiterror policies.

"They're chopping off heads!"

9

"Be Quiet. I Know You Want to, You Know, Save Her."

JUNE SOMETHING, 2016
160-ish Days Until Election Day

I don't know where I am.

I don't know where I am going.

I'm not being coy, I just don't know. I'm always in motion. The other day I woke up, got in the shower, and stood there for what felt like a full two minutes trying to remember where I was. Not just what city, but what state. I eventually remembered, but an hour later I was on the road heading to the airport and I forgot again. Everything looks the same as it does everywhere else, from the highways to the hotels.

One day I woke up and thought I was in London again. Not for a vacation, but *in London*—living there. There was something about the shape of the room, or it was just too dark to see. But for whatever reason, I thought I was home—as if the Trump campaign had never happened. Then I realized I was actually in some Hilton Garden Inn somewhere in the middle of America, and I had to remind myself of what I had been doing for the last ten months. Literally, step by step, campaign stop by campaign stop, I had to rebuild my memory.

Think about that for a second. I had to remind myself of breaking up with my boyfriend, leaving my home, and following around a man some call a maniac, trying to keep up with the daily lies and outrages. He lies a lot. That sounds overly negative. But you have to understand something. Most people, even those who would qualify as political junkies, have other things going on in their lives. They follow politics—but they also go to work, pick up their kids, exercise, shop for groceries, daydream, and live a full life.

I do not.

I live the Trump campaign. That means I live every lie. And I live every controversy. And they pile up daily. Every time Trump opens his mouth in public, I get a verbatim log of it sent to my e-mail. I know him better than I know myself. I hear his voice, not mine, inside my head. For Chrissake, I talk like him.

I have the best words.

I also have a problem.

It's June, and Trump is the presumptive Republican nominee for president. He earned the title unofficially in the first week of May, when Ted Cruz and John Kasich suspended their campaigns. Then he earned it officially on

May 26, when he won enough delegates to clinch the nomination. The party of Lincoln is now the party of Trump.

That's not the problem. I don't root for Trump, and I don't root against him, either. My problem is this: What happens if Trump wins the presidency? Like a lot of political reporters, I don't vote, because I think it's fairer that way. We are not a part of the campaign; we are observers of it. But that doesn't mean political reporters aren't poised to benefit if the candidate they cover goes all the way to the White House.

That's a reality of beat reporting. When the people, places, and businesses you know well do well themselves, you're in demand. If they're a big deal, your work is a big deal. If they take off, your career can take off, too. This is especially true if you have not only knowledge but access.

I've been thinking a lot about access lately. Access is seductive. Access means good nuggets from a campaign. Access means your calls answered. Access is safe and secure, because you're the one at your organization who can always get a comment, a confirmation, or an exclusive interview. But access journalism is barely journalism. And somewhere along the way, out here on the trail, wherever it is I am right now, I decided that access journalism isn't worth it.

D ONALD TRUMP DOES NOT HAVE A CAMPAIGN.
The story goes live on the NBC and MSNBC Web sites on June 6. The GOP nominating convention starts July 18, and Trump needs to pivot from the primary to the

general election campaign. The problem is that Trump is a candidate without a campaign, and my story, reported with my colleagues Benjy Sarlin and Ali Vitali, is the first to punch that point.

We detail a bare-bones team, debilitated by infighting, poor coordination with allies, and a message that changes with Trump's whims. We call attention to the fact that Trump has no communications team to deal with the hundreds of media outlets covering the race, no rapid response director to rebut attacks, and a limited, undisciplined cast of surrogates to launch new ones. This is in contrast to the Clinton campaign, which has over a dozen senior staff dedicated to communications alone. And it matters, because a good gauge of how a person will run the White House is how he or she runs a campaign.

Will they hire the best people?

Will they be organized?

Will they be plagued by scandal?

Will they be transparent?

And so on.

The story catches fire, crackling through social media, and I'm on *Morning Joe* to discuss it at dawn. The story is solid and well sourced, and we expect to see it quickly matched by our competitors at other outlets. But even before that can happen, Trump decides to take us on, and me in particular, doing so personally. If he wanted to underscore our reporting that his campaign is a one-man band, well, he does it in the seven o'clock hour of *Fox & Friends*.

"Katy Tur," he says, calling in, "she knows nothing

about my campaign, she said things about my campaign like she's an expert. We don't even let her in. We don't talk to her. We don't let people talk to her. Because she's a, you know, not a very good reporter."

He also took to Twitter, writing "People like @KatyTurNBC report on my campaign, but have zero access."

"They say what they want," he added, "without any knowledge."

He's trying to discredit me, to intimidate us by insult, but our reporting is good. Do I doubt myself for a second? Of course I do, but then I realize that Trump is not responding to the substance of our story. He is not, say, introducing us to his previously unseen communications team. He's doing his own rapid response rebuttal, for god's sake.

If he wants to tweet, I can tweet, too.

"Why is Trump calling me out on TV/Twitter?" I write, sending a new link to the article and setting off a new round of reactions. Even when I got a note from Hope Hicks, one of the few people doing comms for Trump, it wasn't Hope hitting back—it was Trump, attacking one of my sources.

"Mr. Trump told me to tell you," Hope wrote, "no such 'aide' exists. I dare you to name them."

By the end of the night, Trump is quiet again.

Trump's supporters are never quiet. They also play a lot dirtier. They call me ugly and dumb. They accuse me of sleeping my way to my job. They go after my family, and especially my father, who is transgender. They call me a "cunt." They threaten my life.

The longer this campaign goes on, the more I expect them to take their online rage into the real world. Donald Trump's rallies are dangerous. He likes to say they're "the safest place in our country." He likes to cite "the love in the room." But there's a dark edge to his timing. Often he says this just as a protester is getting forcibly removed.

At Trump rallies, I had gotten used to the screaming and the name-calling. I was a bit jaded about the violent flare-ups— some punches here, some kicks there. We reported every incident. But as it kept happening, it became a bit like sleeping in New York City. You get used to the noise, and then it takes something truly extraordinary to jolt you awake.

That "something" happened for me, and the rest of the Trump press corps, in New Orleans in early March. It was a bad crowd. One man hoisted a KKK FOR TRUMP sign. Another was wearing a Confederate flag shirt. Protesters had a big night, too. They interrupted Trump's rally more than a half dozen times. The largest and longest protest came from a group of Black Lives Matter demonstrators, about twenty of them. They stood up midrally, put their hands in the air, and refused to move.

I looked over from my live position, inside a press pen about ten feet away from the group. Jeers and boos filled the air, rebounding off the endless aluminum and concrete of the venue, an aircraft hangar that suddenly felt like the inside of a ringing eardrum. As the protesters chanted "Black lives matter," the crowd seemed to circle and seethe, chant-ing "All lives matter." With every second the temperature rose and the tension mounted.

Then someone fell, or someone was pushed. Either way the clump of protesters, supporters, and security lurched toward the press barricades. For a moment, they seemed on the verge of toppling the gates, a moment that flipped my stomach and filled my mind with thoughts of a worst-case scenario.

"Louisiana, I am very surprised it takes this long," Trump said from the stage. "That's not like you, that's not your reputation."

Was he calling for a swifter and more extreme ejection?

When the protesters were finally removed, Trump con-tinued to mull the beauty of the old days. "See, in the old days, it wouldn't take so long, folks," he said. "We're in a little bit of a different world today."

That world continued to erupt into violence throughout March, April, and May.

In San Diego, anti-Trump protesters rushed a police barricade. In Albuquerque, they lit fires in the street, forc-ing police to detonate smoke bombs to disperse the crowd. In San Jose, they pelted Trump supporters with eggs and burned Trump hats and other campaign paraphernalia. And in Chicago, the protesters won. The Trump campaign can-celed a rally, allegedly because police told Trump himself it was too dangerous—though the Chicago Police Depart-ment denied it.

At the same time, Trump supporters took extreme countermeasures, ejecting protesters inside of rallies and hanging around for ugly, sometimes bloody confrontations after the fact. In Cleveland, one Trump supporter told a black woman to "go back to Africa." Another supporter yelled, "Go back to Auschwitz." In Tucson, Arizona, a black Trump supporter viciously punched and stomped on a white protester who was with a woman wearing a KKK-type hood. In Fayetteville, North Carolina, a Trump supporter sucker punched a protester, popping him in full view of the television cameras. "The next time we see him," he told a news crew moments later, "we might have to kill him."

Trump himself never explicitly condoned the violence at his rallies, but he never condemned it, either. Instead he seemed to encourage it, like an indulgent father who would never ground his son because of a justified fight. At one point he daydreamed onstage about how much he'd like to punch one particular protester in the mouth. At another point, he promised to pay the legal fees of a supporter who actually did punch a protester in the face. Just about every day for weeks at a time, he ejected protesters with a growl of "Get 'em out of here."

Now, in June, some of the violence has subsided, but Trump has long forecasted "riots" in Cleveland if he is somehow denied the nomination next month. That's not a likely outcome, but it's not impossible, either. Trump is not cooling things down with his rhetoric. Instead he praises his supporters as "spirited," condemns the system as "rigged," and talks ominously of a "rough" month ahead.

———

There's nothing anyone can do," a senior Republican source told me, as though I were the crazy person for asking. "Maybe a priest who can pray?"

I don't know if the priest didn't come through, or didn't try, but Trump is getting wilder the deeper we get into June. The first swing of his general election campaign includes a two-rally Saturday, with stops in Tampa, Florida, and Moon Township, Pennsylvania, where I'm waiting to pick up the afternoon and evening live shots.

That means I'm tracking his speeches, following his remarks. The Republican Party can't change Trump, but maybe he'll change on his own. Maybe he'll moderate what is politely referred to as his "temperament" and more plainly known as the way he is.

Personally, I don't see it happening. I'd love it if it did happen. It would be nice to cover policy instead of personality, but campaign manager Corey Lewandowski has made a strategy out of "letting Trump be Trump."

The philosophy dates back to at least the summer before this one. I remember watching the first Republican debate, where Trump was, well, Trump. Back then it was still a shock to the senses. So during the debate, I sent Corey a text, asking him about Trump's performance, which seemed to have the grace of a falling piano.

"What do you think?" I wrote.

"Trump is Trump," Corey replied.

On this Saturday in June, the statement still holds.

At the first rally in Tampa, Trump steps back from the podium as the crowd chants "Build the wall!" He soaks up the energy of the moment, basks in it, as he usually does. But he doesn't return to the podium; he backs up farther and then drifts toward the right-hand side of the stage, aimlessly at first and then with a sudden sense of direction.

He's heading toward one of the flags. It's like he's drawn to it unconsciously, compelled by a patriotism he can't control. And he doesn't stop. He keeps narrowing the distance, then leans in and wraps the flag in a full embrace. And then he kisses it! He doesn't ask. He just kisses.

Twitter is a bumped hive of activity. Most people have never seen a nearly seventy-year-old man hug and kiss the American flag. I have. I saw Trump hug a flag last August in Derry, New Hampshire. Still, this is the hug heard 'round the Internet. My phone is exploding with notifications. The reactions range from "that poor flag" to "he loves his country, unlike those who burn the flag."

The hug seems to affect Trump like a spice, not a tonic. The rest of his speech—and his entire day—is markedly free of anything resembling a warm embrace. He calls Hillary Clinton a "disgrace," Mitt Romney a "choke artist," and Democratic senator Elizabeth Warren "Pocahontas." It's a reference to Warren's distant and questionable claim to Native American ancestry. Trump talks about how he has been asked to apologize for using the term.

"I said yes, I'll apologize," he says. "To Pocahontas I will apologize, because Pocahontas is insulted."

I s this real life?"

CNN's Jim Acosta isn't sure as our buggy hugs a grassy trail on Scotland's east coast. It's a crisp Saturday in late June, a breeze is blowing, and we can smell the ocean. All our friends are here, but no one has a live shot or an immediate deadline. For a moment, the presidential election is a memory, a childhood game in a town we all left behind.

As we turn a bend, the path straightens and I can see a white MAKE AMERICA GREAT AGAIN hat on the head of the man in the lead buggy. We're at the Trump International Golf Links, in Aberdeen, and the aspiring Republican leader of the free world is giving fifty political journalists a tour.

A cynic might say that this is a self-serving move by Mr. Trump, an effort to showcase his property and burnish his political brand at the same time. A cynic would be on to something. This is the ninth or tenth Trump property that candidate Trump has decided to visit with journalists in tow—and the second Trump Scottish golf course we've seen in as many days. The other was on the western shore, in Turnberry.

Today, we met him on Aberdeen's tenth hole, where he emerged from a white helicopter splashed with the word TRUMP. No tie. Open jacket. A bounce in his step from more than the turf under his loafers.

Early yesterday morning the world learned that England had voted to leave the European Union. Trump reveled in the news and immediately got himself in trouble. Was he concerned about the potential for chaos in the global markets? A crashing British pound? The effect on retirement savings in the United States? On the contrary. Brexit was good for his business.

"When the pound goes down, more people are coming to Turnberry, frankly," he said.

Today, perhaps to prevent more talk like that, the executive vice president of Trump's Aberdeen golf course said the candidate would absolutely not be taking questions. His staff allowed that he might answer a few, but only on the topic of golf.

Trump had other plans, which was good, since I don't know a thing about golf. He practically bounded over to us the moment he landed, and so began our impromptu Wes Anderson frolic through the dunes.

We roll from hole to hole, up dunes, down dunes, between dunes. I zip and unzip my coat as the clouds break and re-form. It is as if Trump is Steve Zissou and we are his idiosyncratic multinational crew. Except instead of an aquatic adventure in a submarine, this is a presidential adventure in a golf buggy—The Life Politic with Donald Trump.

On the thirteenth hole, I follow up on a question I asked yesterday. Maybe by now Trump has found some time to consult with his foreign policy advisers about Brexit and its global implications. He tells me he's always talking to them, but then he shifts into reverse and shrugs off the whole concept of advisers.

"The advice has to come from me," he lectures. "The whole world is blowing up. These people don't have it. Honestly, most of them are no good. Let's go to the fourteenth!"

We go to the fourteenth!

More dunes. Trump tells us his course has the "largest" and "most beautiful" dunes in the world. I can't fact-check that, but it is beautiful here, looking out onto the North Sea, breathing the cool salty air. I let my mind roll out to sea—but Trump notices that Yahoo's Holly Bailey and I are not following him.

"Get up here!" he yells.

It occurs to me that Trump in Scotland is a little different than Trump in, say, New Orleans. He's still bombastic, but he's not encouraging his golf buddies or business partners to forcibly drag any of us off the course. He's saying "Get up here," not "Get 'em out of here." I have a new appreciation for the difference.

"Nobody's ever given you a backdrop like this, ever before," he says, as we look out on the water.

This is the same person who, weeks earlier, trashed me on social media, who slimed me in an appearance on Fox News. I am not surprised. I've come to understand Trump's swinging moods—his shifting respect for me and my job. Yes, respect. No, not always publicly. But multiple people close to him have told me that he does respect what I do. Even when he criticizes it, he also knows the truth.

Take his comment about my having "no access" to the campaign. No, Trump is not publicly praising me, the way he does some Fox anchors or the conspiracy theorist Alex

Jones. He isn't sending me clips of himself with his golden signature, as he does with some folks. But his aides have been as talkative as ever, thanks to constantly warring factions. When one aide doesn't like another, it means everyone likes a reporter.

Possibly Trump is being nicer because he fired Corey right before we left. He likes a good sacking, and while the sacking came as a surprise to Washington, people close to Trump told me the frustration had been building for months. (They said Lewandowski wasn't professional, he didn't get along with the press, and although the Michelle Fields battery charge was ultimately dropped, Trump's kids never warmed back up to him.) Or it could be that Trump is happy to be overseas and away from the drama of the trail. That certainly has much of the press corps feeling lighter.

On the way to the eighteenth hole, I am laughing again. CNN's Jeremy Diamond and Ali are canoodling in a buggy behind me. They weren't good at hiding it in the first place. We all knew something was up when Ali, a Hilton points person, was suddenly staying at a Marriott. Then they were always sitting next to each other, and Jenna Johnson swore she caught them making out in a car garage sometime back in February. They denied it. We didn't believe them.

So far, they're the only Trump campaign reporters shacking up on the road. Or the only ones bad at hiding it.

I look out ahead. The sea grass is blowing. It reminds me of Trump's hair when it's surprised by a strong gust of wind. I haven't had anything but coffee and water to drink, but I feel a bit buzzed. Instead of following Trump up to the green, I walk out to a tee just beneath and take it all in. A bonus moment in the country I just gave up.

Yesterday, back in Turnberry, he made a promise to me. My crew and I were alone on the links, finishing up a couple of live shots, when Trump appeared with an entourage. He waved me over, past Secret Service, and introduced me to his business partners.

"This is Katy Tur," he said. "She is a great reporter."

Come again?

Seeing my expression, Trump caught himself.

"Sometimes," he added.

"You should do another sit-down with me," I told him, seizing the opportunity.

"Yes, I should. I think people would really respect that," he replied. "Okay," he said. "Tell Hope I said yes; she'll set it up."

Business done, he wanted to gossip.

"Do you know who I'm having dinner with tonight?" Trump asked, his eyes squinting, his lips curling into a grin. I'd been given a tip about it, but I wasn't about to let on.

"No, who?"

"You know," he prodded.

"No, Mr. Trump, I don't."

He didn't speak the name. Instead he closed his eyes, pursed his lips, and seductively mouthed two syllables. "Rupert."

He was referring, of course, to Rupert Murdoch, whose media empire includes Fox News.

"Congratulations," I offered, not sure what else I could say.

"Okay, Katy, tell Hope to set up the interview. But only if you promise to be fair."

He rolled off.

I crossed my fingers.

With Corey gone, maybe Hope will say yes. I've always liked Hope. Her inbox must look like an overturned garbage can on a windy day, which is why I have a lot of respect for her work. Most people would be flailing, but Hope is always calm. And she almost always responds.

She gets a bad rap because she's pretty and quiet, and jealous people think the worst.

I find her by the clubhouse in Aberdeen and mention Trump's offer.

"Can we find a time today?" I suggest.

She shakes her head, and we agree to look ahead to times when he's back in America.

So I don't prepare for a Trump interview and, anyway, I've already asked him everything I can think of on the tenth, thirteenth, and fourteenth holes.

I start to drift again—wondering how I'll explain all this in a minute and fifty seconds on *Nightly News* tonight. Should I call it Trump's wild ride? Can I add a Mark Mothersbaugh score? I hang back and watch as the other reporters ask the Muslim Ban question over and over, trying to get some clarity from Trump on what exactly he means. Everyone's always looking for clarity. Trump never provides it.

Then I hear my name.

I should've expected this.

"Katy, come over here. Let's do that interview. Where's your camera?" Trump doesn't so much ask as direct.

I walk over. Secret Service keeps the other reporters back as I try to come up with a new spate of questions. I ask about UK prime minister David Cameron, who stepped down yesterday after the Brexit decision. Trump says he resigned "somewhat in disgrace" and that "he shouldn't have done it."

What about Angela Merkel? Some say Germany is the real superpower nowadays in Europe, I say. Will you make Germany your first call as president?

"We'll see what happens," he demurs. "The world is falling apart."

The other reporters are getting antsy. They're not pleased with the solo joint just out of their earshot.

They start complaining, loudly.

I try to get one more answer before Trump's attention span dwindles.

You haven't been complimentary of Merkel—going after her refugee policy and saying she ruined Germany, I say. Do you think you'll have a better relationship with Russia than Germany?

He gives me his stock "I'll get along with everyone" before cutting me off. "You done? Good."

I walk back to my cameraman and shrug. There wasn't much there, but he did say Cameron resigned in disgrace. That's a bit of news.

Except no, it's not. My microphone wasn't working.

I'm annoyed. I object to Hope that this wasn't really the sit-down he promised, but she doesn't see the difference.

Despite it all, Trump is now officially the Republican nominee. Not the presumptive nominee. The nominee-nominee. His first order of business? Freezing his press corps into submission. What else could he be doing to us in this tundra-turned-ballroom? It's thirty degrees below zero in here.

By here, I mean Doral, Trump's porticoed golf course in South Florida. I'm deciding how to order my questions, wondering if I should start by asking him if he thinks the Geneva Conventions are out of date, when the man of the hour walks in.

He is not smiling. He takes hold of each side of his podium, breathes in, and lets fly.

"So, it's been 235 days since crooked Hillary Clinton has had a press conference," he says slowly, looking at each of the reporters in the room one by one. "And you as reporters, who give her all of these glowing reports, should ask yourselves why. And I'll tell you why. Because despite the nice platitudes, she's been a mess."

He blames her for ISIS, domestic crime, and not having enough American flags onstage at her convention.

"They don't have an American flag on the dais until we started complaining," he says pleadingly. "Then they ran up with two very small little flags, one that we saw."

He accuses her of rigging the system and playing by her

own rules. His big case in point is the private e-mail server she used as secretary of state—a server missing thirty-three thousand deleted e-mails, according to officials.

Trump is asked about the suspicion that Russia hacked into the DNC. On *Morning Joe*, Clinton's campaign manager, Robby Mook, reiterated that the available evidence points to Russia being behind the hack. Two private security firms came to that conclusion. President Obama said that he would not be surprised if it were Russia, given their history of cyberattacks. But yesterday, in a tweet, Trump tried to downplay the claims that Russia was trying to help him.

"In order to try and deflect the horror and stupidity of the WikiLeaks disaster, the Dems said maybe it is Russia dealing with Trump. Crazy!" Trump tweeted. "For the record, I have ZERO investments in Russia."

Now he's hitting that same point on camera.

"It's just a total deflection, this whole thing with Russia," he says.

He's asked about his tax returns and whether they would show investments in Russia or from Russian interests.

"I will tell you right now, zero, I have nothing to do with Russia," Trump says.

He's asked about his relationship with Putin.

"I never met Putin, I don't know who Putin is. He said one nice thing about me. He said I'm a genius. I said, thank you very much to the newspaper and that was the end of it. I never met Putin," he says.

And then he takes an odd turn.

Instead of condemning the DNC hack, or at least saying he needs more information, he looks right into the cameras and makes a strange plea:

"Russia," he says, as cameras click and whir, "if you're listening, I hope you're able to find the thirty thousand e-mails that are missing. I think you will probably be rewarded mightily by our press. Let's see if that happens. That'll be next."

A feeling of disbelief fills the room. Here is a presidential nominee appearing to ask a foreign government to illegally pry into the e-mail server of a private citizen. But Trump is already on to the next topic. "Yes, sir," he says, calling on the next reporter. For some reason the next reporter moves on to another subject. And so does the next. And the next.

I'm sitting here raising my hand, waving it around, fuming. But Trump is ignoring me.

Why aren't they asking about Russia?

I know why. We all have different editors. We all have different questions, different stories. So sometimes a comment doesn't get a proper follow-up, and a thread gets lost.

Another reporter, another question. This goes on for so long I wonder if maybe I missed something. Maybe the Russia stuff isn't news. *Ask your Geneva question*, I think. He keeps saying he believes in waterboarding and torture; it will be a big deal if he says he wouldn't abide by the rules of war.

Trump looks at me.

"Do you think the Geneva Conventions are out of date?"

I immediately regret my question.

"I think everything is out of date," he says. He goes on to say that he would renegotiate it, just like he's going to renegotiate everything, and then moves on.

Dammit, Katy. Now you have to get him to call on you again.

Another reporter asks about Russia but doesn't get to the heart of Trump's earlier comment and why he thinks it's okay to ask Russia to hack into Clinton's e-mail. A few minutes later my patience is up. I just start yelling out my question, hoping to drown everyone else out.

"Mr. Trump, do you have any qualms about asking a foreign government—Russia, China, anybody—to interfere, to hack into a system of anybody's in this country?"

He doesn't answer.

So I try again.

And again.

And again.

"If they have them, they have them," he says.

"Does that not give you pause?" I say.

"You know what gives me more pause? That a person in our government, crooked Hillary Clinton—" Trump says.

I interrupt to get him back on topic. Trump shoots me down.

"Here's what gives me pause," he continues, and I try again to get back to the point.

And he snaps.

"Be quiet," he says. "I know you want to, you know, save her."

Did he just shush me? The whole exchange was a matter of seconds, but did he just shush me?

My phone jumps with Twitter notifications.

"Did he just tell @KatyTurNBC to 'be quiet'?"

"Trump tells @KatyTurNBC to 'be quiet.' "

At the end of the presser, MSNBC anchor Tamron Hall comes back to me for reaction.

"You have tough skin," she says, "but on the day after we saw a 102-year-old woman on the floor applaud the first female nominated to a major party, there you are being told to be quiet by the nominee."

"Tamron, don't worry about me. I'm not good at being quiet, and I won't continue to be quiet in any way, shape, or form," I say.

And it's true.

I kept asking Trump the question. I kept asking until he finally answered.

"If Russia or China or any other country has those e-mails," Trump said, "I mean, to be honest with you, I'd love to see them."

Throughout the month, Trump rallies only get rowdier. I now have private security at all of them. Everyone covering Trump at NBC is under armed protection outside the venues. Inside, where Secret Service prohibits firearms, the security team covers us with a watchful eye and a firm hand. They are themselves mostly ex–Secret Service, in plain clothes, trained to scan for threats and quietly defuse them. They watch our backs during live shots, help clear our paths during entry and exit, and when someone gets too close they step in to keep them away.

But these days, it's not just Trump supporters that sometimes worry them. It's fans of NBC, MSNBC, or my work in particular. Almost all are lovely, politically engaged people of the sort you are happy to meet. But after months of nonstop television, I am getting recognized more on the street, stopped in airports and restaurants. Millions more people than usual are seeing me on *Meet the Press* or *Charlie Rose*, or catching my reporting when it is picked up on competing cable channels and the late shows. The recognition is a thrill. But television's unique ability to make people feel a connection can also be a threat, the security team warns me. What happens if one or two of those people act on their feelings?

One day, delivered to 30 Rock, I get a dozen red roses and a card that reads only, "There will be a day."

I have no idea who sent it.

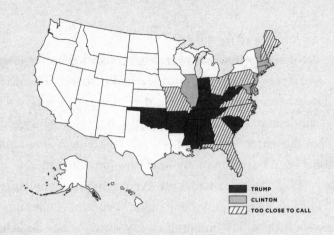

TRUMP
CLINTON
TOO CLOSE TO CALL

TRUMP VICTORY PARTY

NEW YORK HILTON MIDTOWN
8:30 P.M., Election Day

It's going to come down to the wire," an Ohio GOP source is telling me on the phone. "I don't think we're going to get results tonight. But keep an eye on Mahoning County."

I hang up the phone. I must look tired, because our intern, Courtney, asks if I want a coffee.

"Yes, please. And can you bring some cookies?"

Lester Holt is about to come to me live for an update on Ohio. I put down my notepad and check my makeup. The dark circles under my eyes are starting to show. My left eyelid is hanging a little heavier than the right.

Maybe the coffee will help.

No matter how you're feeling, you can never let it show on camera. That's a lesson I learned early on. There were days when I knew one of my older colleagues was pissed off, sleep deprived, or sick. But on TV, he was always golden. "No one cares," he would tell me. "The news is not about *you*." I can hear him in my head now, prodding me.

"Hey there, Lester," I say, trying to pick up my energy. "They do have a bit of good news," I continue, opening my eyes a little wider, pumping up my voice. "They do believe he can make some good numbers up in two counties."

The ballroom at Trump's victory party is starting to fill up. Men in black and blue suits. Women in red and black cocktail dresses. It's a mélange of real Manhattan money and wannabe Wall Street bros. Fox is playing on big screens next to the stage. The crowds can't hear me and they can't hear Lester. But they can hear Fox analyst Charles Krauthammer, and they don't like it.

He's telling Bill O'Reilly in a satellite interview that Trump's isolationist trade policies are just not what the country wants. Trump's nationalist rhetoric worked for him during the primaries, but Krauthammer doesn't believe it expanded Trump's base beyond that. O'Reilly, who begged Trump not to cancel on a Fox debate last winter by trying to bribe him with milkshakes, seems to be regarding Krauthammer with suspicion. As if the longtime conservative commentator might be a turncoat.

"You going to make a prediction that Clinton is the next president?" O'Reilly asks.

"If you force me, I'll take it," says Krauthammer.

The ballroom lets out a great big "boo."

No shade to Krauthammer—he's been doing this a long time—but I don't know if he really gets it, either. I'm not sure anyone in the cable news or network TV studios do. Everyone has been predicting tonight would be a Clinton landslide. But a lot of the places that should've gone to her early are still too close to call. In Ohio, Trump's early ballot totals have already topped Mitt Romney's in 2012. I'm trying to refocus on the NBC broadcast in my ear when Anthony nudges me.

"Look at the screen," he says, pointing at Fox News. "Florida!"

Anthony loves this. For him, it's the Super Bowl, World Series, and NBA Finals all rolled into one. The more drama, the better. Ninety-one percent of the state's vote is in and counted, according to the Fox graphic, and Trump is up. He's beating Clinton by .08 percentage points: 48.9 percent to 48.1 percent.

A couple of hours ago staffers were writing the state off. Now it looks like they might win it. I start imagining the call in my head. Florida goes to Trump. Then Ohio. Then North Carolina. Then . . .

But Chuck Todd cuts me off before I can get carried away.

"Why is the Clinton campaign feeling good and the Trump campaign feeling not so good?"

He's not talking to me, rather the panel on the small NBC screen below my camera, but his question pumps the brakes in my brain.

"It's because of what's remaining, what's out," he says, answering his own question. "Gainesville still hasn't come in, that's a college town."

He's listing all the liberal-leaning Florida population centers, which easily could put Clinton back on top.

"Almost half of the Broward vote hasn't come in, half of Palm Beach, and still about 15 percent of Miami-Dade. And as we showed you, she's winning Miami-Dade by numbers that are landslide-like for Miami-Dade," Chuck says to Lester, Nicolle Wallace, Savannah Guthrie, and Tom Brokaw. They've all seen this show before.

"For me, it's déjà vu," Brokaw says, making the table laugh.

"Florida. Florida. Florida."

10

"Donald Trump Doesn't Lose."

AUGUST 25, 2016
75 Days Until Election Day

We call it "the van," and if I'm not at a rally or press conference, "the van" is where I live. Not down by the river, but at the curb outside of Trump Tower in midtown Manhattan. Calling it "the van" might make it sound like either a bullet-shaped VW from the 1970s or one of those blocky A-Team vehicles. It's more of an industrial minivan with the back end ripped out, replaced with a bunch of TV junk: dials, microphones, video screens, and a microwave dish (not the kitchen appliance) mounted on top of a ten-foot-tall pole that rises from the roof like a submarine's periscope. Even after ten years in TV I can't tell you exactly how it all works. There's a truck operator in charge of the technical stuff, and, in the better vans, anyway, there's a little workspace, or at least a roomy driver's seat.

That's where I'm sitting in late August when I become suddenly aware of a figure outside the window.

One of the reasons we spend a lot of time in the van is because we're parked on Fifth Avenue. There's a mad, muddy Mississippi River of people and vehicles coursing by every second. In fact, it's less as if we're in a vehicle and more as if we're in the attic of a house, riding out a terrible storm. On the other side of the van's windows there are loud noises, terrible smells, a lot of nice, kind tourists, and more than a few folks in the mood for politics. Some people take selfies, with looks of genuine pleasure. Others direct a middle finger at the tower, with looks of genuine pain. There's a woman with a small dog who does eight-hour stretches with a WOMEN FOR TRUMP sign and, with equal commitment, a man dressed as a giant penis. His head is his head. His balls are his feet. He offers PHOTOS WITH DONALD TRUMP.

The guy outside our window looks more normal than most. He's older and kindly looking, with a gray brush mustache and a newsboy cap. I'm reminded of my grandfather, my mother's father, the Greek pool builder who liked to dance. So, okay, what the hell?

I pop open the door and say hello. The street dust stings my nose and burns my eyes. Traffic sounds pound my ears. And the thick August air, my God. It pushes into the cab like a wet tongue. But I'm doing this. I'm meeting the public. That's not meant to sound bratty or sarcastic. I believe in viewer feedback, no matter how harsh, as long as it's constructive.

I also believe that the more viewers have a chance to talk with us, the more they know us and the more power we have as reporters. We can tell the truth all day, but it's pointless if no one believes us. What journalists need most of all is trust, and trust is only built through relationships.

The older gentleman says hello back and shakes my hand.

"How are you?" I say. "What's going on?"

Cable is different from network news. Both breed intimacy, because both come right into your home, often during dinner or breakfast or some other moment when domestic life is happening and attendance is invite-only. But network morning and evening shows are only on during particular windows, and there's a lot less audience loyalty than you might imagine. Every day, or almost every day, you're addressing millions of people you're meeting for what might as well be the first time.

That's not true on cable. Cable is on all day. Viewers know you like they know their own family. When you're doing live shots every hour, your personality bleeds through. There is nowhere to hide and, even if there were, you'd be too tired to hide there.

I always knew this intellectually, but the last few months have made it real. Just a couple of weeks ago, Trump was in Des Moines for a midafternoon rally. Around lunchtime my friend Ashley Parker, from the *New York Times*, got an e-mail from a woman who lives just a few blocks from the arena where Trump was speaking.

"I sure wish I could treat you, Sopan Deb, Maggie, Katy

and Jenna to ice cream today," the woman wrote, in perhaps my favorite example of fan mail. "I follow all of you, and love thinking about you being in the neighborhood today."

She signed it, "all the best, and ice cream too."

This older gentleman at the van door does not seem interested in buying me ice cream. He's saying something about Trump and my coverage, but I can't hear him over the traffic noise.

"Excuse me," I say again. "I'm sorry. The traffic is loud."

"I said, I work in this building here," he says, or something like that. I still can't hear him, so I tip my head even farther, so my whole right ear is facing him.

That's when he starts yelling.

There are polls out there, from Gallup and Pew, with very depressing things to report about the news media and public trust. As an industry, we've been on a four-decade slide down from a high point shortly after Richard Nixon resigned.

One of the theories blames Republican politicians, heirs to Nixon. It's true that they've spent years attacking and undermining the press. But that theory has a fatal error. You see, throughout the 1970s, the percentage of Americans who reported a "great deal" of trust in the news media was already abysmal—somewhere between 18 and 21 percent.

I have my own theory. I think it has something to do with our extreme discomfort with conflict. That might sound odd at first. Americans love conflict, right? It's the very soul of democracy—a contest of ideas. Or that's the big, beautiful idea, anyway.

I think we secretly hate it. I think we dislike and ultimately distrust the media because journalism, honestly pursued, is difficult and uncomfortable. It tells us things about the world that we'd rather not know; it reveals aspects of people that aren't always flattering. But rather than deal with journalism, we despise journalism. And in the case of this older gentleman, we end up yelling about it.

"You female reporters are so obtuse," he screams, and this time I definitely hear him, although I don't know what he means.

I do understand the particular antipathy a person might feel toward a reporter who is covering a politician they love. After all, politicians spend millions of dollars trying to make millions of "friends." Their whole careers depend on convincing you that they know you, care for you, and are defending your interests. After Franklin Delano Roosevelt died, a story circulated about a common man who collapsed in grief at the sight of the president's funeral procession. A neighbor picked the man up.

"Did you know the president?" he said.

"No," the man said. "But he knew me."

Now, that story is probably apocryphal. But it shows the kind of love that's out there for some politicians, and Trump is, for a good portion of America, a politician who inspires that kind of love. "Nothing short of Trump shooting my daughter in the street and my grandchildren" can dissuade me from voting for Trump, a woman told Ashley Parker of the *New York Times*.

So imagine how you would feel if every time you turned

on NBC, you saw my reporting on this figure you love—this figure you think will lift you up, save your job, make your country great again. Imagine how you'd feel if every night and all day this little blond-haired girl was shining a critical light on your beloved figure. Who is she to question his plans? Double-check his statements? Follow up on his promises?

You would hate me.

And people do.

People like this older man.

His anger isn't words anymore; it's a growl and then something guttural, now on the tip of his tongue and then . . . patoo-ey. There are only inches between his face and mine, but the spit seems to float between us for hours and, besides, thoughts move fast. Here are mine:

Since his convention in Cleveland, in July, Trump has crashed through the guardrails of traditional politics. He has feuded with the family of a soldier killed in Iraq, invited Russian hackers to meddle in American politics, declined to endorse House Speaker Paul Ryan, appeared to joke about gun lovers assassinating Hillary Clinton, and called President Obama "the founder of ISIS."

He's also done a bunch of little boneheaded things. He questioned Senator (and former POW) John McCain's dedication to veterans, defended his year-old musings on which orifice Megyn Kelly's "blood" flowed through, and linked Senator Ted Cruz's father to the Kennedy assassination, among other lowlights.

Oh, and he blew up his staff again. Breitbart chairman Steve Bannon and pollster Kellyanne Conway are now in senior positions. Paul Manafort—who himself had replaced former campaign manager Corey Lewandowski—is gone, resigned under a cloud of questions regarding his ties to a pro-Putin leader in Ukraine.

Trump trails by an average of eight points in recent nationwide polls.

At some of his rallies, it's as if he doesn't even care anymore.

"I just keep doing the same thing I'm doing right now," he'll say, looking out at his supporters, some of them passed out in the summer heat. "At the end it's either gonna work or I'm going to have a very, very nice long vacation."

Is he really that casual about all this? There are tens of millions of dollars and just as many dreams yoked into his gambit of a run for president—and he's going to shrug about the possible outcomes?

The old man's spit breaks up in a gust of wind, but my face is covered in a fine mist of strange saliva. It occurs to me that I just took a loogie for a candidate who may not even care if he becomes president. Then I think, *Trump's got the right approach.* I'll keep doing what I'm doing—reporting, writing, questioning. At the end, politics is either going to be my permanent beat or I'm going to have a very, very nice long vacation.

Actually, let's hope for both.

Covering a political campaign turns out to be a cross between getting married and starting a new job. As with any new job, you begin wearing your best clothes with your best face forward. But over the months, high heels and pressed shirts give way to flats and tees. We've all started dressing like the Saturdays we rarely get off anymore. Jeans of any color, style, or cleanliness are fine because no one is going to see them, anyway. On TV, you're only shot from the waist up. My motto: Just add a blazer! It's the TV version of the mullet. Professional up top, party down below.

Some aspects of our appearance can't be helped. Like chipper new husbands and wives, we committed ourselves, but the long hours, bleak work, and spotty appreciation have taken their toll. We don't look after ourselves the way we used to, and virtually all of us are fatter, balder, and paler than we were at this time last year.

Not for lack of effort. A couple of weeks ago, on a Sunday, I was home in New York. I promised myself I'd get to a yoga class. This in itself was a break from my now-usual home routine, which revolves around trips to Whole Foods (apt nickname: Whole Paycheck), and weak-willed splurges on unnecessary new garments and shoes. Researchers say that when people are tired they are more likely to overeat and shop to excess. Well, duh. I buy the expensive food to feel renewed on the inside, after too many bags of in-flight peanuts and sad hotel salads. I buy the $450 pair of black patent-leather loafers because what's on the outside is an issue, too.

The retail therapy never works. There is no replacement for rest and relaxation, the old R&R. On this particular Sunday, I was trying to recover another way, the real way: exercise. I not only promised myself a yoga class, but yanked on some spandex and made it to the studio on time. I stuffed my keys and wallet into a little cubby, then did something out of character with my phone. I shut it off.

For the next fifty-five minutes, I stretched, bent, and oooooooohhhhhhmmmmm'd. Then I picked up my phone and my keys, got a cup of water, and stepped outside. Maybe it was an exercise high, but I felt amazing. Then I turned on my phone—and felt like hell. Eight urgent e-mails. Ten voice mails. Twelve text messages. All with some variation of "WHERE THE F ARE YOU???" Clinton had apparently had a medical incident at a September 11 memorial ceremony, and our whole team was thrown into action. Except me, because I was busy bending and stretching. The one hour in a whole goddamn year I turned off my phone. I haven't shut it off since.

Today we're packed inside an airplane hangar. It's late September in Melbourne, Florida, a day after the first presidential debate, and we're waiting for Trump's first post-debate rally. The atmosphere is like the molten inside of a microwave burrito, complete with a dirty metal sleeve to help radiate the heat in every direction. Trump's supporters are fired up, as usual, but I can't help but notice they're also a bit wilted.

People are using giant Trump signs and MAKE AMERICA GREAT AGAIN hats to fan themselves. Those hats, my goodness: NBC first reported them on sale at Trump Tower on July 25, 2015, sometime around 5 P.M. They sold out within half an hour. Now practically every Trump supporter has one. Today, the effect is like red sprinkles atop a great melting mass of vanilla ice cream.

Last night's debate still has me grinding my teeth—not because of what happened onstage, although Trump flung enough half-truths and falsehoods to tire even the most diligent watchdog. I'm frustrated by the way the assignments were handed out. I woke up to discover that, although NBC News was host of the debate, and my badge says I work for NBC News, I wasn't a core part of the coverage on the network side.

Lester Holt was the moderator, Chuck Todd and Savannah Guthrie were the live analysts and hosts, but I didn't have a camera with me on the floor of the spin room. In other words, there was no expectation that I would be the one talking to Donald Trump when he came off the debate stage. It meant that although I was a major driver of news about the Trump campaign, I wasn't going to have much of a chance to drive it on NBC's big night.

I tried not to take it personally. I went for a little walk. I talked it through with Anthony. I called some friends. Still upset, I called my bosses to find out what was going on and how I could contribute to the night. To their credit, they talked me through it. But I probably sounded a little more annoyed, maybe even crazed, than I should have, and now, twelve hours later, the feeling hasn't exactly worn off. To be fair, I take everything a little too personally nowadays. We all do. The competition to get on air is so much tougher than it was in the beginning. There are only two candidates left and only one that anyone seems to want to talk about: Trump.

It gets so intense among us that we clash over the dumbest shit: who's getting invited to campaign-organized conference calls, or who confirmed some inconsequential piece of logistics first, or who got ten extra seconds on this show or that.

The morning Exchange has started to feel more like the Hunger Games than a companywide editorial call. We jump all over each other to try to convince the bosses that we have the best analysis, or the newest source reporting, or are in the most ideal location when the shows air.

Either NBC lucked out or NBC News and MSNBC chairman Andy Lack is a genius, because we somehow have the hardest-working and, in my opinion, best TV political team covering the election. Kasie, Hallie, Kristen, Jacob Soboroff, and I are getting rave reviews for consistently breaking news, consistently offering deep analysis, and consistently providing the most well-rounded coverage. NBC seized on it and branded us the Road Warriors because we are constantly on the road. They ordered up a big promotional push, with hot photo spreads in glossy magazines like *Rolling Stone* and *Esquire*. And when we happen to be in the same place at the same time, we get an afternoon or late-night show on MSNBC—a loose roundtable about what it's like to be on the trail.

Like most wonderful things, though, it is a blessing and curse. We genuinely like each other, but we genuinely want to beat the hell out of each other, too. You don't get this far if you aren't deeply competitive. So in the moment everything feels so important. So immediate. So crucial. Sometimes we say things we don't mean. Things that keep us up at night. Things that make us nauseous and guilty. Things we hope we'll forget on November 9.

It's not just the reporters. The producers are territorial and the embeds are suspicious. Everyone is eyeing everyone else, regardless of what news network or paper is listed on their press badge.

"How did *she* get *that* interview?"

"She sucked up."

"Did you see her Twitter feed?"

"Not one critical comment."

"It's shameful."

Or:

"Can you believe *he* peddled that reporting off as *his* own?"

"He's always *seconding* my e-mails."

"I heard that, too. I heard that, too."

"If you heard it, too, why didn't you say it first?"

And on and on and on.

Right now, in this blazing heat, I'm succumbing to it again. My frustration builds with every drop of sweat.

I didn't even have a camera!

Drip.

Have they seen how many hits my tweet has gotten?

Drip.

My reporting was picked up in the Washington Post*!*

Drip.

Despite the lack of resources in the spin room, I managed to push Trump on a big unanswered question. During the debate, Hillary Clinton had gone after Trump over his tax returns, which he has declined to release, breaking with decades of precedent.

"Maybe he doesn't want the American people—all of you watching tonight—to know that he's paid nothing in federal taxes," Clinton said. She also mentioned the one tax return of Trump's that people have seen. It shows him writing off major losses, which could theoretically allow him to duck federal taxes for years. Onstage, Trump made it sound as though he was defending that very fact, saying, "That makes me smart."

But what's the bottom line? Does he pay taxes or not? That struck me as a simple, important question the public should know, so I decided to corner him on it. My producer and I fired up our iPhones and hid behind another reporter as Trump drifted down the rope line. I was afraid he might want to avoid me. But when I popped up—"Mr. Trump!" I said—he actually smiled.

"Hi, Katy," he said, and then he finished another answer about online polls, which, of course, showed him winning.

"A quick question about taxes, Mr. Trump," I said, and the smile vanished.

We bantered a bit about whether the government wastes tax money, and then I got to the point.

"Do you pay federal income tax now?" I said.

"My current returns will be released as soon as they . . ." he began.

"Do you pay federal income tax now?" I persisted.

Trump walked away.

The clip had a couple of thousand retweets within hours and got picked up by other outlets.

But I'm still feeling raw here in Melbourne.

It doesn't help that I'm wearing wool pants, because I woke up in New York and it was cold there, and because I'm an idiot. I filed for *Nightly News*, a piece about Trump's past comments about a former Miss Universe's weight. Now I'm just waiting for my live on-camera close.

The close is a tradition from the early days of television, a fifteen-second stand-up at the end of a cut piece. The point is to show the audience that you're a firsthand authority on your subject. Typically, I love a good on-cam close. But tonight, I'm dragging. My hair looks like a half-wrung mop. Anthony is back at the gate, pleading with Secret Service. They confiscated my dry shampoo—no aerosol.

Down in the crowd, a man with a long, greasy pony-tail is walking back and forth, hollering, hoping to start a "lock her up" chant. I always wondered how those chants got started. Tonight the crowd isn't biting. Too hot. Trump's not here yet. The night is young.

Anthony emerges from the crowd and he brings forth a miracle: he got my dry shampoo past the Secret Service. Don't ask, he says. I spray it on and comb it through with just minutes before air. Then I square up to the camera and prepare for my tag.

The subject is Trump's relationship with Alicia Mach-ado, the 1996 Miss Universe winner. Trump, who co-owned the Miss Universe Organization until last year, called her "an eating machine." She says he also called her "Ms. Piggy";

Hillary Clinton brought all this up at last night's debate, and Trump didn't deny it. In an interview with Fox that morning, in fact, he explained that she was far from the ideal winner of a beauty pageant.

"She gained a massive amount of weight," he said, "and it was a real problem."

You know what's evidently not a real problem for Trump voters? Any of this. That's my tag.

"Tonight Trump is here in Melbourne, Florida, where voters who support him—female voters who support him— say they don't care what he says because he used to be an entertainer. And, yes, they do want a female president. They just don't want it to be Clinton."

D inner is eight donut holes. Eight delicious donut holes that put me right to sleep. I get about an hour in before I force myself to wake up. I feel bad about leaving Anthony to drive all alone. We are heading back to New York from New Hampshire, and as midnight rolls around, I refresh my e-mail—and there's nothing. No one is asking me to be on television tomorrow. Not. A. Single. Show. In fact, a note from one of the senior producers in D.C. tells me I've got the whole day to sit in 30 Rock and make some reporting calls. I am practically giddy at the news. Yes, I can appreciate the irony: after years of fighting to get on the air, all you want is one day off it. Tomorrow is my day. *Hell, I might even try yoga again.*

Less than eight hours later, my phone is buzzing on the

nightstand. I ignore it, but it buzzes again. And then again. And then again. The clock reads 7:37 A.M., aka normal people time, but instead of a leisurely breakfast, followed by some reporting calls, I am being summoned to 30 Rock for an 8:55 A.M. live shot. I think about ignoring the e-mails and turning off my ringer. But I also get notifications when Donald Trump tweets something, and that's when I see what's going on.

He's been up since 3:20 A.M., firing off tweets about Alicia Machado, the former Miss Universe. She's been calling Trump unfit for the presidency, her remarks amplified by the Clinton campaign. But really, Mr. Trump? It's September 30, four days after the presidential debate, three days after my *Nightly News* spot. The main ring of American politics has moved on from Alicia Machado. Why are you pulling her back into the conversation? And ruining my day in the process?

I speed walk the half mile from my corporate apartment (my friend finally got sick of climbing over my stuff to get to the bathroom and kicked me out of his apartment) to 30 Rock and start commenting on what I know about the unknowable. Sometimes it feels like my bosses think that in my spare time, I am Trump's therapist, listening to him detail his every state of mind or urge, in weekly sessions. Why would Donald Trump do this thirty-nine days before a presidential election? What is he thinking?

The truth is, I have no idea. He was doing so well, as far as any of my sources were concerned. He was reading the prompter. Avoiding petty fights. Playing nice. Staying on message. Steve Bannon and Kellyanne Conway, or perhaps RNC chairman Reince Priebus, had reined him in, tamed him.

Of course, this is relative. Trump is still Trump.

He is still picking up whatever grenade Hillary Clinton throws at him and tossing it back, no matter how nonsensical. No, you're the bigot! No, you're the one who founded the "birther" movement! No, you're the liar! No, you're the one with the bad temperament! No, you're the puppet! No, you're the one who is bad for women and minorities!

But these tweets are a return to the bad old days.

In the 5 A.M. hour, Trump called Machado a "con," the "worst," and "disgusting." Oh, and he invited his millions of followers to "Check out" her "sex tape."

By midday, I am no closer to understanding, and I say so on air. I keep thinking about a meeting I had a few weeks earlier, with a source very close to Trump. We sat in a diner, not long after Trump's early-August meltdown, and this source was adamant: Trump has a chance, a good chance. He claimed that the candidate was a new man, recommitted to winning.

"He doesn't lose," this source kept saying. "He doesn't lose."

He kept repeating this like the words to a charm.

"Donald Trump doesn't lose."

Donald Trump has been endorsed by zero current or former U.S. presidents; zero Fortune 100 CEOs; and only one newspaper—the *National Enquirer*, aka Elvis Lives.

All true.

But Donald Trump doesn't lose.

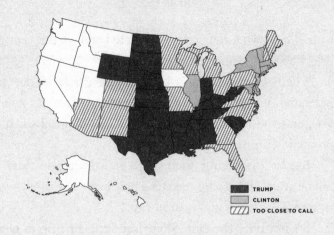

TRUMP VICTORY PARTY

NEW YORK HILTON MIDTOWN
9 P.M., Election Day

I feel a tap at my ankle and look down. Ben Jacobs is below me.

"Do you have a charger?" he says, like Oliver Twist asking for some more porridge.

A few hundred miles away, my friend Phil Rucker feels something similar—a figurative tap on his shoulder.

"I think I should peel off," says Phil Rucker.

He's talking to Karen Tumulty at the *Washington Post*. They're in the *Post*'s newsroom in D.C., writing tomorrow's front-page election story, the one for posterity as much as the present. Until now the story was about how Clinton had won. That's what the polling predicted and what recent political history seemed to assure.

The *Post* is relying on what everyone is relying on: state polls, national polls, internal campaign polls, history, analysis, and the behavior of the candidates themselves. It all points to Clinton.

But Phil is having doubts. "Keep working on this," he tells Karen. He's going to start a new draft of history—one in which Trump wins—just in case.

It's not only the *Post*. Story lines are shifting in newsrooms across the country. Hours of precut Clinton packages are now in peril, and, as a Trump presidency seems more possible by the minute, senior producers and editors across the country are scrambling the jets.

Ben is still looking at me.

I glare back as if to say, "Where's your fucking charger?" This is Election Night. Are you kidding me, Ben Jacobs?

I toss him my cord and he plugs it in to one of the four thousand power strips on the riser. Then, without asking, he climbs onto the riser and sits cross-legged at my feet. Have I mentioned there is no space?

But I like Ben, so I don't tell him to scram. He's the only person in the traveling Trump press corps who's been on this ride as long as I have, although I don't remember seeing him in the beginning. At that point he must have had his own charger. He says he was there in New Hampshire at the backyard pool event in June 2015, which means he's seen Trump rip the wig off George Washington and give the Founding Fathers an atomic wedgie.

I think the prospect of Trump actually winning might make Ben a little ill. Not in a political sense, but in a which-way-is-up sense. Or better yet, in a can-I-be-sure-of-anything sense. He's a twenty-first-century René Descartes, wondering if anything he sees, feels, smells, or hears is real. Do facts exist? Does reality exist? Or is this all a dream?

I let him stay at my feet, but we barely talk. No one is really talking on the press riser. We're alone together, staring off into space. Meanwhile, the crowd is getting drunker and rowdier. Fox News calls Kansas, North Dakota, South Dakota, Texas, and Wyoming for Trump. Virginia, Michigan, Pennsylvania, Colorado, and Arizona still aren't in. But the race is a toss-up, a fact that's moving to the front of the house. It's there on the face of Fox's Chris Wallace, there in the way he fidgets, there in his eyebrows. A mischievous, almost devilish thought is coming.

"I think we're all—" he starts, then pauses, searching for the right way to put it. "At least I'm coming to the conclusion tonight—conclusion is the wrong word, open to the possibility, Donald Trump could be the next president of the United States."

The ballroom goes absolutely nuts, and the TV feed is interrupted by cheers from a crowd amassed on the street outside the Fox News studio.

Wallace gets sheepish, adding: "Folks, I just said it was a possibility!"

I look over at Fox's Carl Cameron, my neighbor here on the riser. He's sitting on his chair, half facing me, chewing nicotine gum. He has the stare of a man who believes he has seen this show before and if anyone has, it would be him. He covered his first presidential race in 1988. I want to know what he thinks, and he knows I want to know. I don't even need to open my mouth.

"It's still early," he says, matter-of-factly. "The urban vote is gonna kill him."

He says the words *kill him* with his whole mouth, then looks back at his camera.

I turn to NBC's live coverage, which plays on a personal monitor that our crew has screwed to a pole beneath my camera. Lester is about to toss to me, but something about the exit polls catches his eye.

"Can we go back to Michigan?" he says, asking the control room to restore a full-screen graphic showing an outline of the state. "I'm not sure I saw it correctly. Was that a double-digit difference?" The graphic pops on-screen: Trump's face on the left, Clinton's face on the right, and in between them a seventy-one-vote difference—just double digits.

Lester is awed.

11

"Grab 'em by the Pussy."

OCTOBER 9, 2016
30 Days Until Election Day

Traditional Republicans never liked Trump, didn't want him as a candidate. They accepted him because of the overwhelming will of his voters. But their support was not unconditional. Two days ago it appeared that Trump had violated a key, if unspoken, tenet of Republicanism: thou shalt not brag on tape about grabbing pussy.

Now they want out. Not even Trump can survive this.

"This is the end," a GOP source is telling me on the phone. It's 11:30 A.M. on a Sunday, one month before the election, and I'm sitting in the lobby of Trump Tower. One phone is smooshed between my ear and my shoulder. I'm taking notes on my other phone and also trying to keep an eye on the elevators, watching as the golden doors open and close, waiting for anyone associated with the Trump campaign to come down. They're going to have to show their faces sooner rather than later. The next presidential debate is in St. Louis tonight.

My source is ranting, but Trump Tower is oddly quiet. Most of the other Trump reporters are already in Missouri or back at the Essex House, which has become the official hotel for out-of-town Trump reporters. As far as I can tell I am all alone, except for the occasional tourist seeking refuge from the storm outside. A hurricane is rolling up the mid-Atlantic, but I'm told it is nothing compared with what's brewing sixty-six floors above me.

Across the country, Republicans are abandoning their nominee in droves, rescinding their endorsements, and condemning his candidacy. They question his moral character. They call his comments "predatory," "abhorrent," and "offensive." They say his campaign "has been nothing but a race to the bottom" and it is now "impossible" to support him. For weeks the Clinton campaign and Clinton supporters had been airing audio and video of Donald Trump saying derogatory, objectifying things about women.

"A person who's flat-chested is very hard to be a ten."

"You know, it doesn't really matter what they write, as long as you've got a young and beautiful piece of ass."

"Does she have a good body? No. Does she have a fat ass? Absolutely."

Through all this, many Republicans were silent. But now all are suddenly realizing that, surprise, they know a woman. And are now horrified on their mothers', daughters', sisters', and wives' behalf. Republican Florida congressman

Tom Rooney went so far as to say he would be failing as a father if he voted for Trump. It would be like "telling my boys that I think it is okay to treat women like objects," he said.

My source is a GOP state chair in a swing state, meaning her state is purple, a toss-up between Republicans and Democrats. She is an important person in an important state who is supposed to be organizing the effort to elect Trump along with the rest of the Republicans down the ballot. The tape doesn't help anyone, but it's putting the most stress on Republican senators in tight races. Right now she is worried. "Every party in every state is preparing for the Democrat ad that tries to pin Trump's groping comments to their vulnerable GOP senators or congressman," she says. It's not that Trump's language is not consistent with the Republican brand; it's not consistent with the brand of a twenty-first-century human. "Any Republican serious about winning doesn't stand up for this," she says. But it's more than that; she's now questioning whether she, herself, needs to quit the party. "How do I maintain my integrity and do the job I was elected to do? When you hold an elected position you do have an obligation. How do I do my job and not turn into a horrible human being?" she says. The question is still hanging in the air when we hang up.

Then it's just me on a hard black bench, staring at bright golden elevator doors, waiting for a Trump staffer to provide a reaction to all this. Is he going to drop out? What does he think of all the Republicans calling for him to drop out? How is he going to go forward? Will he be contrite at the debate, or will he continue to be Trump? What is the plan? Is there a plan? Is he actually sorry? Or is he just sorry he was caught?

As I run through a mental list of questions for Trump, one comes to mind for rank-and-file Republicans: Why drop Trump now? Why is *this* the line for Republicans? Why not calling Mexicans rapists? Or fighting with the pope? Or racism toward a federal judge? Or the fraud lawsuits? Or the name-calling? Or the fearmongering? Or the xenophobia? Or the countless other degrading statements he's made about women, including his own wife? For god's sake, why not the half decade of birtherism?

You remember that, right? Trump's one-man crusade to prove, without evidence, that the first African American president was illegitimate because he was born in Kenya. Last month on *Hardball*, former Georgia Republican congressman and current Trump surrogate Jack Kingston tried to claim that birtherism had nothing to do with Trump's rise. I couldn't believe my ears. This guy, who I had never seen at a Trump rally, was trying to tell me—on TV—what Trump supporters thought. It was ludicrous, and when he finally stopped interrupting me, I told him so.

"Let me tell you what Donald Trump supporters tell me on the campaign trail," I began, trying to raise my voice above his. "Let me tell you what Donald Trump supporters tell me on the campaign trail often. They often say they believe he was born in Kenya. They often say they believe he's a Muslim. Some of them go on to say that they believe he's an undercover operative, a Manchurian candidate, if you will, that has the interests of a foreign power rather than the interests of the American public."

No, the reality is that Trump's base liked the comments coming out of their candidate, and Republicans didn't drop him because that base was growing and growing. I mean, it's not as though Trump had apologized for the birtherism or even acknowledged it as a theory that he had, if not hatched, then at least nurtured, groomed, and paraded down Main Street.

It's easy to say that Trump's campaign is over, but hard to imagine him dropping out or even apologizing, given the way he ultimately dealt with the birtherism. Trump refused to back away from it until it was absolutely politically necessary for him to back away from it: when polls showed 44 percent of the country's likely voters thought he was a racist and only 1 percent of African American voters said they would vote for him. This was just three weeks ago.

He held a big to-do at his brand-new D.C. hotel, luring our cameras with the promise of a come-clean news conference where he'd finally surrender to the truth and say what everyone knew all along: the president of the United States was born in the United States. We got there, set up, and went live, all of us. CNN. MSNBC. Fox News. Trump took the stage alongside his family and—bizarrely, given the advertised topic—a dozen or so veterans.

"Nice hotel," Trump said, opening up the event. "Under budget and ahead of schedule."

There were chuckles and toothy smiles from his family

onstage. But the press was not amused. We were not there to cover a hotel. After duly plugging his property, Trump handed off to the vets, who gave aimless soliloquies about why they think Trump is such a fantastic guy. After twenty minutes or so of live coverage, all three networks cut away. It was clear that we had been duped and that Trump was trying to shield himself in the service of others.

But after another interval, Trump reappeared at the microphone. At last, he seemed ready to get the whole birther thing over with. But instead of apologizing, he kept lying.

"Hillary Clinton and her campaign of 2008 started the birther controversy," he said, although she did not.

"I finished it. I finished it," he said, although he did not.

"You know what I mean," he added, although we did not.

And then he finally said it: "President Barack Obama was born in the United States, period. Now, we all want to get back to making America strong and great again. Thank you, thank you very much."

Seven seconds. Seven measly seconds. That's how long it took Trump to dismiss a lie he told for five years. He did not apologize. He did not take questions. It was as if his years of exploiting a divisive, racist conspiracy theory were secondary to a hotel tour—which, to him, they probably were.

The stage collapsed a couple of minutes after Trump left it. It literally buckled and fell while I was on live television. I heard it happening, looked behind me, paused, looked back at the camera, rolled my eyes, and continued. Trump had rick-rolled the press, lied to retract a lie. Reality could no longer support itself.

For months Ben Jacobs has been saying—only half-jokingly—that nothing matters. Trump has done dozens of things that were once considered political suicide. He has survived them all. But will he survive this? Will Trump's recorded talk of a pussy grab matter? Will these last-minute Republican defections matter?

A campaign staffer walks in from the rain and, to my surprise, sits down right next to me, across from the Trump Tower elevators. He looks like a man who is thinking about his next job. I ask him what's going on upstairs. Headquarters is basically empty, he tells me. He's barely seen any of the core campaign staffers since the news broke on Friday. Kellyanne Conway, Trump's normally ubiquitous campaign manager and the first woman to run a major-party presidential bid, is MIA. Reince Priebus, the head of the Republican National Committee, is privately telling folks in the party that it's okay to quit if they're worried about their political future.

"Why such a strong reaction?" I ask my source.

"The guy has been in the public eye for thirty years," the source says of Trump. "There's got to be more."

The *Access Hollywood* tape, as it's now being called, since *pussy-grab* tape isn't as TV friendly, broke two days ago, on Friday afternoon. I was called into NBC's executive wing around 3:30 P.M. "You're not in trouble," the executive said over the phone. "Just get down here."

She was in what looked like a pretty serious phone conversation when I walked into her office. She pointed to her computer, hit play, and told me to listen. It was video of an *Access Hollywood* bus driving toward the camera on a movie lot. Since her speakers were crap, I had to all but press my ear into the monitor to hear the audio. I had no idea what I was supposed to be listening for.

The video was of the bus, but I could make out two disembodied male voices, one of which I immediately recognized.

"I did try and fuck her. She was married," said the voice, the same voice I had heard almost every hour of every day for the past fourteen months.

I looked over at the executive.

"Is that Trump?"

"Yes," she said, barely turning from her phone call. "Keep listening."

"I moved on her like a bitch. But I couldn't get there," I heard Trump say. "And she was married. Then all of a sudden I see her, she's now got the big phony tits and everything. She's totally changed her look."

Who is Trump talking about?

"Sheesh, your girl's hot as shit. In the purple," said the other man.

Who is Trump talking to?

"Whoa! Whoa!" said the other man. "Yes! The Donald has scored. Whoa, my man!"

Trump and his gentleman friend were going back and

forth about someone off camera. A woman who clearly couldn't hear them.

"I gotta use some Tic Tacs just in case I start kissing her," Trump said. "You know, I'm automatically attracted to beautiful—I just start kissing them. It's like a magnet. Just kiss. I don't even wait. And when you're a star, they let you do it. You can do anything."

"Whatever you want," mimed the second man.

"Grab 'em by the pussy. You can do anything," Trump added.

Oh. My. God.

My head was cymbals clashing, sirens flashing, trains crashing.

"Did Trump just say he can grab them by the pussy?!" I yelled, suddenly unconscious of my surroundings.

The executive got off her call and said yes. Trump did. She had been on the phone with the head of NBC News Standards, the department that decides what is suitable for air and what is not, particularly when the situation is hairy.

"Who is the second voice?" I asked.

"Billy Bush."

Oh.

The tape was from 2005. *Access Hollywood* had it in their archives and the *Washington Post* somehow got hold of it. Because *Access Hollywood* is an NBC Universal property, the *Post* called us for comment, I was told. "Go find an office and get me a script," the executive said. "We go to air as soon as we can get the video into the system and subtitled. Legal and standards has to review what you report before you report it. Reach out to the campaign. We need their response. Do it quickly."

I walked out and found an empty office. It would be helpful if I had my own desk somewhere in the building, but since I've been on the road no one has quite figured out where I belong while I'm in 30 Rock. So I sat down, uninvited, and opened my laptop. The first thing I did was send the campaign what has to be the most bizarre request for comment and context I have ever written.

Hope and Jason,

NBC News has obtained a video of Mr. Trump during an interview with *Access Hollywood* from 10 years ago. Mr. Trump is on a hot mic having a conversation with Billy Bush bragging about hitting on a married woman.

"I did try and fuck her. She was married."

He also talks about how he behaves with women he is attracted to. He says, "When you're a star they let you do it. You can do anything. Grab them by the pussy."

The video is from 2005.

The *Washington Post* is dropping this in about an hour. We will be following suit with the video.

Does the campaign have a response or context?

Thank you,

Katy

About an hour later, after legal and standards reviewed my every word, I was on the air describing the video and, in the kind of disclaimer typically reserved for crime and crash videos, warning that the content was graphic and potentially offensive.

The trickiest part of the script was finding TV-safe phrases for Trump's language. I felt like the dub team trying to get an NC-17 movie into shape for a basic cable audience. Instead of *try and fuck her*, I used *try to have sex with her*. That was easy enough. But *grab 'em by the pussy* was trickier. I settled on *sexual advances*, which, in retrospect, was unfair to sexual advances. It's not as though Trump was talking about a soft touch of the hand at dinner.

Network news, even cable news, is particular about what it allows on the air. In most cases, we blur out blood, we freeze the frame before a gun fires, we certainly don't say "pussy." There are social standards.

The story dropped like a cannonball into a kiddie pool. In an instant, the pool was broken, the jumper was hurt, and everyone around him was soaked. In this case, that meant the Republican Party. Trump didn't have a sterling reputation when it came to women before this. So the comments, bragging about sexual assault, being able to do whatever he wants because he is a star, were, for many, a mere confirmation of his true character.

That Friday night Trump apologized. I remember turning on the TV in my corporate apartment, leftover takeout boxes on the table, multiple suitcases open and half-unpacked on the floor. Trump had posted a video to Facebook, and it was virtually simulcast on every network.

"Anyone who knows me knows these words don't reflect who I am," the candidate said. "I was wrong, and I apologize."

That was the first time in the entire campaign I had heard him say he was sorry. But it wasn't a stand-alone apology. Trump had been backed into a corner. The campaign had forced him to record something, and I thought I could see how angry he was to have to do it. He added a twist. Just as he had with his birtherism retraction, he used the moment to attack and threaten the Clintons, the better to take the heat off himself.

"I've said some foolish things, but there's a big difference between the words and actions of other people," he said. "Bill Clinton has actually abused women and Hillary has bullied, attacked, shamed, and intimidated his victims. We will discuss this more in the coming days. See you at the debate on Sunday."

For Democrats the attempt at equivalence was infuriating. For Republicans it was frustrating and humiliating. Newt Gingrich and others had tried and failed to take Clinton down for his infidelities in the 1990s. Now Trump was using the same playbook, except he's not running against Bill Clinton. He's running against Hillary. A woman.

The "we will discuss this more in the coming days" promise is on my mind when the elevator doors open and Trump's new campaign CEO, Steve Bannon, walks out looking no more rumpled or haphazard than usual. Bannon has only been on the team for two months, but he's already become something of a legend. Not a legend in a Babe Ruth or Mickey Mantle sort of way. More like the Sith or Voldemort, according to critics. He's got a long and confusing résumé, from naval officer to movie producer and investment banker, but most notably, before he signed on with Trump, as the head of Breitbart. The right-wing "news" outlet is infamous for headlines under Bannon's tenure, such as BIRTH CONTROL MAKES WOMEN UNATTRACTIVE AND CRAZY; HOIST IT HIGH AND PROUD: THE CONFEDERATE FLAG PROCLAIMS A GLORIOUS HERITAGE; and BILL KRISTOL: REPUBLICAN SPOILER, RENEGADE JEW.

"Steve!" I say, getting up to catch him before he rushes out of the lobby. He's in a good mood, or at least he's projecting one. "Is Trump going to go after Bill Clinton tonight? Is he going to bring up his accusers?"

Bannon smiles.

"Wait and see," he says, and disappears behind a service door held open for him by Secret Service.

My day began with a source telling me the Trump campaign is over. By the looks of Bannon's smile, though, it's not over yet.

Trump ditched us. His plane to St. Louis took off from NYC before the campaign plane carrying me and the rest of the press. Have I mentioned that we have a campaign plane? We do. We got it after the convention. I'm not on it very much. MSNBC wants a reporter on live TV at all times—so most times I'm traveling ahead or behind on a commercial flight, but it's fun when I can make it.

A press plane is a typical part of late-season campaign coverage, because the campaigns kick into such high gear that it's impossible to keep up using commercial flights alone. What's also typical is that the candidate would travel on the same plane as the reporters who are following him. But Trump has defied that convention, too. He has his own plane, and although it is a 757, there is no room for all the reporters now covering him. (Remember when it was just me? Yeah, me, too.) I don't mean that there are more of us than can fit inside a 757; I mean that Trump's 757 is stuffed with all kinds of other things: two private bedrooms, gilded bathroom, movie room, VIP lounge, dining room, and on and on. It can only carry forty-three passengers—down from a normal 757 capacity of well over two hundred passengers.

So, yes, we travel on our own plane. It's paid for by the news networks, print publications, and wire services that are covering Trump. And because Trump isn't on it, ever, it's easy for him to ditch us. It is commonly thought that Clinton hates the press, too, maybe even more so than Trump, but even she doesn't have the gall to ditch us. Trump has the gall—and he takes a certain joy in it as well.

"I have really good news for you," I remember hearing him tell supporters in Laconia, New Hampshire, after he ditched us last month. "I just heard the press is stuck on their airplane. They can't get here. I love it."

We in the press do not love it. But we deal with it. We also grouse at the ways in which Trump's occasional media blackouts could hurt him in a long-term historical sense. Consider that Trump is already at the debate site when our plane lands in St. Louis. We board a couple of buses. Normally, one bus would be reserved for the "protective pool," which is the smaller group of reporters, cameramen, and photographers assigned in a rotation each week to get very close to the candidate and send pool reports back for use by the entire press corps. These people are not there for gotcha questions or trivial breaches of Trump's privacy. No, this is a team on the history beat. When Trump ditches this pool, he ditches posterity. He runs the risk of some major event occurring in the middle of a media blackout.

I've tried to explain this to folks on Twitter who don't seem to understand why it's important for candidates or presidents to have a protective press pool. We know what George W. Bush did in the moments after the 9/11 attacks because of a pool report. We know what happened to JFK in Dallas because of the pool reporters who were on the scene. We have history books thanks to the work of hundreds of pool reporters over the years, and hundreds of other political beat reporters, too. But people don't seem to appreciate this, not when there is social media available to fill the gap.

Information coming directly from a politician or his team, without being vetted by reporters, is little more than propaganda. No American voter accepts one-sided accounts in their personal life. We wouldn't trust our teenager's perspective on a fender bender. We wouldn't trust a single co-worker's description of a crucial meeting. We wouldn't even wholly trust our best friend's version of a nasty breakup. We look for holes in the story. We look for more information. We should demand the same in politics. And yet so often we do not.

We really have to start teaching journalism in elementary school. People don't even understand the basics of what we do anymore.

Anyway, we pull up to Washington University, where protesters are lining the main road to the debate site. Demonstrations are to be expected. Cheeky signs are par for the course. But in the aftermath of Trump's *Access Hollywood* tape, the anti-Trump brigades have some newfound flair. Many of the signs say the same thing: HANDS OFF MY PUSSY, with *pussy* replaced by a pictogram of a cat.

We all rush to one side of the bus to get a shot of the signs.

And here I thought the anti-Trump WE SHALL OVERCOMB *or* THERE WILL BE HELL TOUPEE *signs were clever.*

It's not that the reporters want pictures of the signs because they are anti-Trump. It's because Trump himself had to drive right by them on his way in. Unless he was closing his eyes, he couldn't miss them.

Inside the auditorium, I set up my laptop in the official filing room, which is next to the debate spin room. There are giant screens in front of me, because I'll be watching the debate on television. Yes, I'll have flown hundreds of miles to attend an event, only to be one or two rooms short of being inside the event itself. The entire media will be watching it on TV, exactly as we could have in our own beds with a pillow propped against the headboard and a pint of

Cherry Garcia in our hand. The reason we have to be there in person is all the pre- and postcoverage, all the run-up and reaction work—which tonight is immense.

The big question in the aftermath of the Trump tape is what else is out there. Are there other people who will come forward with personal accounts of what Trump boasted about on the tape?

At the same time, we are also poised to cover what Trump will do in response.

Less than two hours before Trump and Clinton face off, and just thirty minutes or so after I arrive in St. Louis, we find out the reason why Bannon had been smiling back at Trump Tower. Yes, Trump is indeed going to bring up Clinton's accusers. He's just not going to wait for the debate to begin.

Trump's advance team invites reporters to a nearby hotel and directs anyone who can't make it to a live video feed on the campaign's Facebook page. Since I have predebate live shots, I'm watching the live feed, where Trump appears alongside four instantly recognizable women: Paula Jones, Kathleen Willey, Juanita Broaddrick, and Kathy Shelton. The first three women have been famous since the 1990s, when they each accused Bill Clinton of forcing himself on them. The fourth woman, Kathy, is less well known, but her story is no less serious. She said she was raped at age twelve and that Hillary Clinton defended her alleged rapist in court in 1975.

One by one they read statements defending Trump.

"Actions speak louder than words," says Broaddrick, who

alleges that Bill Clinton raped her in a hotel room while he was attorney general of Arkansas in 1978. Broaddrick didn't come forward until 1999, when Clinton was president. Clinton denied the accusation and was never charged.

"Bill Clinton raped me," she says, drawing a contrast with Trump's mere talk. "Hillary Clinton threatened me," she goes on. "I don't think there's any comparison."

The reporters in the room are shell-shocked. For some reason we continue to be surprised. In the aftermath, it's reported that Trump is planning to put the women in the Trump family box, right up front for the live debate. That would mean Bill Clinton, who has always denied any wrongdoing, would have to come face-to-face with his accusers. It would also mean that when Trump is questioned about the "pussy" tape, which he very obviously will be, he'll have a visual response: four allegedly abused women, something for the people to see, not just hear. The man knows television. No disputing that.

Tonight is do-or-die for Trump. The party is going to decide whether it sticks by him or cuts its rope, largely depending on how he does. And the *Access Hollywood* tape isn't the only thing that's dogging him. The first debate was a disaster for Trump; he was unprepared and easily flustered. Clinton, who had been practicing for weeks, knew how to get under his skin, and she succeeded within minutes.

Instead of brushing off the little digs or relentlessly refocusing on the strength of his economic message, Trump interrupted Clinton anywhere from twenty-eight to fifty-one times, depending on how you were counting. He said "wrong" so many times that a supercut of the performance went viral. At the time it felt like I was watching a ballerina dance around a cranky toddler with a hammer.

As we get closer to the start of this debate, the media room fills to capacity. It seems as if every American journalist is here, along with every journalist from every country on the planet. As befits a serious discussion between candidates vying to become the leader of the free world, there is also beer and animals. Budweiser is the debate food partner, serving up a hot buffet backstage, accessible only by walking through a branded barn with real Clydesdales in it where various VIP GOP donors and guests are taking selfies.

The debate starts.

No handshake.

The first question is about pussy grabbing.

Here we go.

Even the Clydesdales are quiet. Trump calls his comments "locker-room talk" and, yup, tries to refocus on the Clintons. He calls Hillary an enabler who bullied Bill's victims. But he's handicapped. The Commission on Presidential Debates declined to allow the four accusers a spot in Trump's family seats. Good-bye, visual aid.

Trump is also breathing heavily. The microphone picks up every sniff, which makes Trump seem angry. He calls Clinton "the devil." He says she has "tremendous hate in her heart." He argues she should be in jail because of her use of a private e-mail server while she was secretary of state.

The debate's town-hall format is also presenting a new hurdle for Trump. Instead of podiums, there are two stools.

The candidates are encouraged to walk around the circular stage and interact with voters. In truth, this is rather ridiculous. No one looks comfortable sitting on a stool. I understand that the organizers are trying to go for something folksy and intimate, but they might as well require every candidate to be eating an apple as they chat, or working out on an elliptical machine. How about a debate that features each candidate driving home from work with a Bluetooth clipped to his or her ear and questions coming in via the car speakers? *Kids, kids, quiet down. Daddy is debating.*

There is no nonstaged way to be "off the cuff" on a stool or, for that matter, walking around in a circle while a bowl of voters listens to you explain policy. And Trump—well, I don't think Trump has ever sat on a stool. Where would he? He doesn't drink—because, he says, his late brother Fred was an alcoholic—so he probably doesn't go to bars. And I can't imagine him sitting at his kitchen counter eating breakfast with Melania. The one time I did see Trump at a diner, he sat in a booth.

So instead of sitting, Trump is pacing. He's breathing hard *and* he's pacing *and*, from some camera angles, it looks more like he's looming over Clinton, stalking her as she answers her questions. It would have been complicated for any man to run against the first woman nominee of a major political party. But Trump has made it infinitely more so, even before the *Access Hollywood* tape.

For months he has portrayed Clinton as someone who is only successful because she is a girl. He argues that, as a candidate, "the only card she has is the woman's card," by which he apparently means that women seem to like her more than him. This gives her an unfair chance of winning, he argues. In fact, he likes to say at his rallies, she'd have no chance to win if not for "the woman card."

There's a certain amount of obvious strategy here. Trump is working to beat her by neutralizing the thrill of electing the first female president. He's trying to give voters who might like the idea of making history an excuse to instead vote against that history.

But what's interesting—and troubling to Trump's many critics—is that Trump is taking what would be an unquestionably historic achievement and using it against her. He is using her gender against her, seeming to imply that being a girl makes her unfit for the presidency. He says that she doesn't have the "strength or stamina" to be president. Or that she doesn't have the presidential "look." Or that she was "schlonged" by Obama.

His supporters eat this stuff up. In fact, as Trump fights to survive the *Access Hollywood* tape, I keep thinking of what his supporters have been saying for months—via T-shirts and tchotchkes. The first time I saw a TRUMP THAT BITCH shirt was in New Hampshire. This was a long time ago, before the primaries. I was heading into a Trump event when a vendor's display caught my eye, and there it was—in multiple colors, multiple sizes. I thought it was a one-off, the work of a freelance misogynist or, at best, opportunist. Maybe it was. But it spread. And spread.

The first time I saw a HILLARY SUCKS BUT NOT LIKE MONICA T-shirt was back in April in Albany, New York. Again, a vendor was selling it outside of a rally. He asked me if I wanted one. He didn't say it snidely. He didn't even know I was a journalist. No, I told him. I don't want the shirt. It's a disgusting shirt. And yet that shirt became a staple, too. The next one was a pin that caught my eye, the KFC HILLARY SPECIAL, emblazoned with the phrase 2 FAT THIGHS, 2 SMALL BREASTS, LEFT WING. Get it?

Lately, those shirts and pins are not even the worst of it. Just a couple of weeks ago in Melbourne, Florida, a man, with his wife proudly by his side, showed off his own creation. It was a royal-blue tee with white block letters that read I WISH HILLARY MARRIED OJ. He posed for pictures in front of the press pen. Let me say that again. He posed for pictures in front of the national media in a shirt that unsubtly conveyed that he wished Hillary Clinton had been brutally stabbed to death in the 1990s.

If you had asked me a year ago, I would never have thought that Americans would stoop so low or accept gratuitous name-calling—on either side, against either candidate. But now, as Trump sniffs and paces onstage tonight in St. Louis, it suddenly seems easy to imagine him surviving this scandal—or even getting a bump from it. Trump is crude, and in his halo of crudeness other people get to be crude as well.

That may be why Clinton is not spending the whole debate talking about Trump's *Access Hollywood* tape. Instead she keeps bringing up another news story from Friday. It was forgotten a bit in the fever around Trump's video, but it was important, maybe even more important to the future of the country. On any other day, in fact, it would have been the subject of front-page, special-report, nonstop coverage. I'm talking about a joint press release from the Department of Homeland Security and the Office of the Director of National Intelligence. It read in part: "The U.S. Intelligence Community (USIC) is confident that the Russian Government directed the recent compromises of e-mails from US persons and institutions, including from US political organizations."

Translation: You know how the Democratic Party server had been hacked? You know how the e-mails of Clinton campaign chairman John Podesta had been stolen? Well, we think we got the guy. His name is Russia. And we think that Russia is stealing this information with one singular goal: "to interfere with the US election process."

The release did not say that Russia wants Trump to win, but Clinton is making that case onstage in St. Louis. Again, Trump has made it easy for her. His former campaign chairman was paid by a Putin ally. His staff changed the GOP platform to seemingly benefit Russia.

That episode was a particularly strange one. At the Republican convention, Trump campaign staffers got picky about an amendment supporting "lethal defensive weapons" for Ukraine—weapons that might be used to fend off pro-Russia forces. Republican leadership and conservative foreign policy hawks supported the position, but Trump's folks demanded that it be stripped from the platform while leaving everything else intact. Perhaps not coincidentally, Trump's campaign chairman Paul Manafort used to work for a deposed Ukrainian president who supported Putin.

Now Trump won't definitively say he'll come to the aid of any country attacked by Russia. He won't release his taxes even to prove that he has no financial ties to Russia. In July he publicly called on the Russian government to find a tranche of e-mails deleted from Clinton's private server. He's also been saying fawning things about Russian president Vladimir Putin and, with undisguised glee, he's been campaigning on every scandalous and embarrassing e-mail to emerge following the hacks. A new batch landed just a few days ago. Now Clinton is trying to portray Trump as Russia's plaything. A candidate Russia will benefit from. A candidate Russia is actively trying to help.

"She doesn't know if it's the Russians doing the hacking," Trump responds. "Maybe there is no hacking. But they always blame Russia. And the reason they blame Russia is because they think they're trying to tarnish me with Russia."

Trump is vicious, but he is also cooler and calmer than he was during the last debate. He doesn't even come out to the spin room after the debate. Instead, the campaign sends out Clinton's accusers and a couple of staffers including a thirty-one-year-old speechwriter named Stephen Miller.

Tonight, Miller says, was "the greatest debate victory in the history of the United States."

The next morning on our way to the airport, our driver, a local fireman and father of two daughters, tells Anthony and me what a great job he thought Trump did against Clinton.

Is he bothered by the *Access Hollywood* tape? He is not.
"Why?"

"Because Clinton is a liar," he says.

The enthusiasm for Trump isn't waning. The crowd at the Mohegan Sun Arena in Wilkes-Barre, Pennsylvania, the night after the debate, three days after the *Access Hollywood* tape, is just as big as ever and even rowdier than others have been. When I walk in, a man and his wife are leaning over the bicycle racks of the press pen. At first it looks like they're trying to talk to a CNN reporter. When I get closer, though, I realize they're actually yelling at her.

You need more makeup, the husband advises. Keep piling it on! On second thought, don't bother, he adds. You're so ugly it won't help. He is going on and on, heckling this reporter as his wife stands there, laughing. They are trying to break her concentration, to draw a reaction, but the reporter is a rock. They're about four feet from her, but she is completely ignoring them.

I wish I could have done the same. Instead I start yelling at the man. I tell him to stop it and I congratulate his wife on being married to such a great guy. After they walk away, the reporter walks over to her camera. A couple of minutes later she is on live TV. The couple is standing off to the side still staring at her, hoping she'll mess up or give some hint that they rattled her. I'm staring, too, hoping she doesn't give them the pleasure. She doesn't miss a beat.

The thing is, they don't look like cruel people. They look like they're in their forties. They're wearing designer jeans and nice boots. They seem healthy and comfortable, and it's hard to imagine them acting this way at home or in the office. Hillary Clinton recently called Trump's supporters a "basket of deplorables," and while some might be easy to single out like that, most aren't. A lot are your coworkers and your neighbors. They're your taxi driver, your fireman, and your supermarket cashier. They're the mom in riding boots and a Barbour coat helping her cute daughter with her school science project. You would never know that they're Trump supporters, quote unquote deplorables.

But inside a Trump rally, these people are unchained. They can drop their everyday niceties. They can yell and scream and say the things they'd never say out loud on the outside.

"Obama is Muslim!"

"Hillary Clinton is a cunt!"

"Immigrants need to get the hell out!"

"Fuck you, media!"

On the outside, this kind of behavior is disreputable. But inside a Trump rally, they can tell a woman she's ugly and needs more makeup. They aren't deplorables. They are patriots. And now they're waiting to see their hero for the first time after a three-day fight for his political life.

It's just after 7 P.M., and Trump is in the building. The crowd is chanting, "Drop dead, media! Drop dead, media!" They've been waiting in the arena for hours, listening to the same six or seven songs in an earsplitting loop. Every time there's a pause in the music, they take a collective breath and start to cheer, hoping that it means Trump is coming. They're ready for him. I spot a new T-shirt in the crowd:

SHE'S A CUNT, VOTE TRUMP. The guy is with his wife and three kids.

As Trump's campaign has evolved, so have his speeches at the rallies. It's still all about him: his polls, his unparalleled ability to win, his gift for deals. But now he's also portraying himself as patriotic. He used to walk onstage to the song "You're the Best." Lately, he's striding out to "God Bless the USA."

The crowd goes wild.

After the rally, Anthony and I head to a Panera Bread down the street. We've been to this chain a bunch of times before. But tonight something feels off. We're eating our salads when we both look at each other and then look around. It feels like we are being watched, and not in a friendly way. It may be in our heads, maybe we're just paranoid, but we leave our salads and hit the road.

October 12. I'm in Lakeland, Florida. The *Access Hollywood* news is five days old. Trump is holding a rally on the tarmac of a private airport. Outside the airport, a lone protester is playing the tape on repeat. "Grab 'em by the pussy. You can do anything," Trump says over and over.

On the way in, I ask an older woman in a flowery red dress what she thinks of the tape.

She takes an uncomfortable breath and then smiles.

"What person hasn't said it?" she says.

This is not the answer I am expecting.

"I haven't said it," I say.

"Well, good for you," she says.

"You've said you're going to grab women by the—?"

"No. Don't be stupid," she says.

Onstage Trump kisses a WOMEN FOR TRUMP sign. He actually puts his lips on it, presses his face against it. I'm doing a live shot with Bloomberg's Mark Halperin, who notes how much energy the WikiLeaks dumps are giving Trump and Trump's base. The rhetoric is red-hot and so is the tarmac. We're all on it, watching Trump. But while the reporters could bring in umbrellas, Trump's supporters could not. The paramedics are here, too, which is good since lots of people are collapsing. Even so, few are leaving. They want to stay. They want to see history.

A couple of minutes after I finish my *Nightly News* spot I get a tip from a friend. The *New York Times* is going to drop a bombshell any minute. *Okay*, I think, *this is it*. The *Access Hollywood* tape could not possibly be the only shoe to drop. Thirty or so minutes later, as Anthony and I are deciding what hotel to sleep in tonight, our phones start buzzing. The *Times* story has landed. Two women are alleging that Trump groped them: one on an airplane in 1979, the other outside a Trump Tower elevator in 2005.

We decide to book the last flight back to NYC. We get into Newark after midnight. Although my alarm is set for 7:30 A.M., I don't get to sleep that long. The assignment desk wakes me up at 7 A.M. Ivanka Trump is going to Pennsylvania today to rally female voters in Philadelphia's collar counties. NBC wants me to get her on camera reacting to the women who have accused her father of sexual misconduct and to the 2005 audiotape. She's relevant because she is campaigning for her dad, trying to soften his image and convince women he's not as bad as they might think.

It's a long-shot assignment. Ivanka does not "react." I've never seen someone so aware of herself at all times. It's as if she assumes there's always a camera on her—which, to be fair, there usually is. The chance of us getting her to break her silence and actually address reporters is almost nonexistent. But I get dressed as fast as I can and take the train down to Philly. Anthony rents a car and drives down from New Jersey. He picks me up from the station and we drive another forty-five minutes to the first event. We get there five minutes before Ivanka shows up.

CNN has the same idea. They sent Dana Bash. I don't think she was too pleased to see me walking in the door. Not because we don't like each other. We do. But it would have been a lot easier to get Ivanka if only one of us had been there. Her events, today, are billed as "Coffee with Ivanka." She shows up, all smiles, in an A-line blue dress. She doesn't address the headlines: not the tape, not the accusers, not the still-lingering speculation that her father will drop out of the race. She repeats the same lines she's been using since the start of the campaign. Her dad is her "personal mentor," her "role model."

This is almost too rich. I mean, come on. I talk to some of the women in the crowd. Are they buying this? They are. They tell me how impressed they are with Ivanka. They believe her more than they believe the women who have come forward against her dad.

Dana and I tried, but we didn't get Ivanka to talk to us. She didn't even acknowledge our presence as she rushed by us. I called Ivanka to try again. She did answer the phone, I'll give her credit for that, but she declined an interview or even an on-the-record comment.

The next day I'm in Charlotte, North Carolina, sitting in the front seat of a Suburban fitted with a satellite on the roof, trying to watch a news conference. Yet another woman is accusing Donald Trump of forcibly touching her. This time it's a contestant on *The Apprentice*. The monitor playing the presser is behind my head, so I'm twisted around in my seat, balancing my laptop on the center console.

I have a live shot as soon as this finishes. Producers want me to talk about this latest accusation and the other allegations that have continued to surface. It's been a week since the *Access Hollywood* tape dropped, and so far nine women have come forward to say Trump acted inappropriately.

There are the two women from that first *New York Times* piece a few days ago. Jessica Leeds, the woman on the plane, told the paper that Trump "was like an octopus" and that "his hands were everywhere."

The woman waiting for the elevator in that same piece, Rachel Crooks, told the paper that Trump kissed her "directly on the mouth" without asking.

Now there are more. A woman helping a photographer, Mindy McGillivray, told the *Palm Beach Post* that Trump grabbed her ass.

A *People* magazine writer, Natasha Stoynoff, wrote a first-person confession for the magazine about the time Trump pushed her up against a wall in a room at Mar-a-Lago and forced his tongue down her throat. Trump's third wife, Melania, was pregnant at the time.

A former Miss Utah, Temple Taggart McDowell, told NBC News that Trump kissed her on the lips without asking.

A former Miss Washington, Cassandra Searles, wrote on Facebook that Trump grabbed her behind and repeatedly invited her to his hotel room.

Kristin Anderson said Trump reached up her skirt and touched her genitals at a nightclub in the 1990s. "He was so distinctive looking—with the hair and the eyebrows. I mean, nobody else has those eyebrows," she told the *Washington Post*.

A makeup artist, Jill Harth, filed and then withdrew an "attempted rape" lawsuit against Trump in 1997. Despite that, she told the *Guardian* she stands by her claim that Trump tried to force himself on her in Ivanka's bedroom at Mar-a-Lago.

BuzzFeed just published a story claiming Trump walked in on contestants of his teen beauty pageant while they were changing. Some girls, their sources said, were as young as fifteen.

Now Summer Zervos, a former *Apprentice* contestant, is crying and telling the world in a news conference that Trump "thrusted" his "genitals" on her in his Los Angeles hotel room. She said they were supposed to be discussing a job.

The campaign and candidate are having a hard time keeping up with the accusers. Initially they denied the allegations one by one but now they've resorted to a blanket denial of everything and a counterpunch, calling the women opportunists or Democratic operatives. It's not the most persuasive denial ever given, but it doesn't seem to matter. Last night, Ben Jacobs tweeted a picture of a middle-aged woman in black-rimmed glasses at a rally in Columbus, Ohio. She was wearing a white low-cut tank top with a homemade message scrawled on the front in red marker: TRUMP CAN GRAB MY . . .

Below that, she drew a shaky arrow pointed down toward her crotch.

A week ago in Trump Tower a Republican source told me, "It's over."

Maybe so.

I'm just not so sure it's over for Trump.

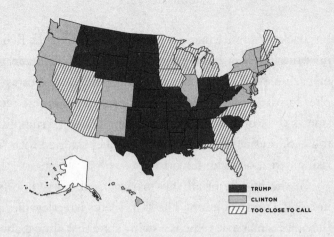

TRUMP VICTORY PARTY

NEW YORK HILTON MIDTOWN
11 P.M., Election Day

Wall Street isn't happy about Trump's chances tonight. At about 10:30 P.M. the Dow fell more than seven hundred points, in futures trading. If that holds when the markets open tomorrow, it will be the biggest point decline since the 2008 financial crisis. And it's not just American markets in free fall. The prospect of a Trump presidency has sent Japan and Hong Kong into a spiral and sunk the value of the Mexican peso to a two-decade low.

It's a strange reaction to the idea of America's first CEO in chief, a candidate who has run on the claim that his business skills will transfer to government.

Back in March a major consulting firm called the Economist Intelligencer Unit, a corporate sibling of *The Economist* magazine, put Donald Trump on a list of looming catastrophes for the world economy. They ranked a Chinese recession as the worst risk, then slotted in "Donald Trump wins the U.S. presidential election" at number six, tied with "the rising threat of jihadi terrorism."

In the middle of all this market freak-out, Ohio falls into the Trump column. The shift seems irretrievable, and the news is massive. Ohio is a swing state that almost always swings toward the victor. I text Sean Spicer, to see if the party is seeing what we're seeing.

"Ohio?" I write.

"Yup," he says.

I'm looking around the ballroom, whose occupants are getting drunker and more jubilant by the minute, but something is weird. There haven't been any chants of "Lock her up! Lock her up!" I open a fresh reporting note for the NBC listserv, our politics e-mail share, noting that tonight feels a little different from a typical Trump rally. It's almost like the responsibility of governing is having a sobering effect that's even stronger than the celebratory cocktails.

Except not. Almost as soon as I send my note I hear the chants. So I send another note to rescind my last note. Then I get on the phone to some of my sources. The Clinton team is telling Andrea Mitchell that Clinton's most likely path is through Michigan, Pennsylvania, and either New Hampshire or Nevada. But is Michigan still in play?

Not according to Trump's people.

"It will be tight, but we will win Michigan," my source tells me. He says it again: "We will win Michigan." It feels like another decade, but it was only yesterday that Trump made the last stop of his campaign in Michigan, in the wee hours of the morning. Tomorrow will be our "Independence Day," he said. "We're hours away from once-in-a-lifetime change."

Now it seems he might have been right. The Obama coalition is not turning out in the same numbers for Clinton. I'm putting this into a reporting note when I'm interrupted by a cranky Paul Revere type who comes striding by the press pen. "I told you so!" he yells. "I told you so!"

When I look back at my phone I notice a picture from Sean Spicer—a shot of the whole Trump team inside "the war room." Sean and some senior aides are in the front, and the camera is aimed over their heads at Kellyanne Conway, Mike Pence, Jared Kushner, Ivanka, the Trump boys, and Donald himself. He's looking up at what I presume to be a monitor showing the results. Trump's campaign team is smiling, but he and his family don't look so happy. Maybe they don't want to jinx it. Maybe the camera caught them all between smiles. Maybe it's something else.

For all his bluster, Trump's own data had him most likely losing. As a result, his campaign spent a lot more time on his concession speech than his victory speech, or at least that's what campaign sources told us. Now they're frantically crafting the words of a president-elect.

"Flor-ri-da!"

The ballroom begins a new chant.

"Flor-ri-da!"

It's the next big prize of the night, and it's shading toward Trump. Trump staffers are hugging one another and whispering, shaking each other's shoulders, slapping each other on the back. It's like they're mission control after a moon landing, and in a sense they are. If Trump goes to the White House, he'll be the first of his kind to do it. The first president with no prior political experience and no prior military experience. He's no Neil Armstrong, but he's a man on the moon.

In Pennsylvania, Clinton is up, but not by as much as Obama was after the big urban areas are counted. The rest of the state is expected to vote heavily for Trump, potentially swamping her lead.

I see a new note in the system: a Trump adviser on the ground in Pennsylvania tells NBC that the state is a "jump ball."

And then another one comes in. We have a reporter in Miami, at a Clinton watch party, where people have been dancing for most of the night. But now they're watching the returns, waiting to see if Florida goes to Trump. No more salsa. The music has died.

12

"*There's Something Happening, Katy.*"

OCTOBER 24, 2016
15 Days Until Election Day

But really now. This campaign is over, right?

But really now. This campaign is over, right?
Whatever wisdom there is at the bottom of a bottle
of wine seems to confirm it. It's after midnight at Bern's
Steak House, a fancy but kitschy restaurant in Tampa. It
boasts "the largest wine cellar in the world"—before a half
dozen or so Trump reporters began our unquenchable assault. How many bottles have we had? No one is sure. We
arrived around 9:30 P.M., after Trump's rally. We had to wait
for a table. Now all the velvet-backed chairs are empty but
ours. We're closing the place down.

And why not? This feels like the end of the line. Trump
trails by double digits in some polls. No candidate has ever
come back from that far behind this late in the campaign. A
lot of Republicans seem to agree. Senator Mike Lee of Utah

has called for Trump to drop out, while John McCain, the party's 2008 presidential nominee, has withdrawn his support. One Trump staffer recently told me that he is so confident he'll be done—job over, final check, good-bye—on Election Day that he already paid for his dream vacation. Hell, even Republican House Speaker Paul Ryan has cut ties with the Trump campaign, an emergency measure meant to save other Republicans running for Congress.

At tonight's rally in Tampa, Trump showed none of this weakness. Speaking beneath a handmade sign emblazoned with a thumbs-down drawing and the word MEDIA, he blasted America's "rigged" system and its "dishonest" press. The ceilings shook with chants of "Lock her up!" but a certain hopelessness and desperation hung on Trump's every word. He seemed to be putting the bitter in bitter end.

So when word spread on the press riser that a bunch of us were going to get dinner, almost everybody said, "Screw it, I'm in." And now here we are, lifting our bottomless glasses, guaranteeing ourselves a short night and queasy live shots in the morning. But it's bloody glorious. Eighteen months ago, I knew a few of these people, mostly by reputation, but I couldn't imagine sitting with them as professional peers. Now, unrelenting as this campaign has been, I'm sad to think that it's almost over.

Think about what we've been through. For the rest of our lives we'll need each other just to vouch for stories that our children, spouses, and other friends surely won't believe. The shock-a-day style of the Trump campaign is overwhelming our means of recording it. Human beings have an edit button in their heads, an amnesia switch that none of us consciously controls.

That's why we forget some of the craziest or most painful parts of our lives. We have to forget them, because it would be impossible to deal with the present if the past were such a second-by-second burden. This quality influences our politics, too. Trump said Clinton had been "schlonged" by Obama, but that can feel like a distant rumor when he's using softer language. Trump pointedly refused to condemn endorsements from a white supremacist and former KKK leader, but that can dissolve into hazy memory when he's speaking with an African American pastor. George Orwell said seeing what's in front of your nose demands a constant struggle. It's also a constant struggle to recall what's in the back of your mind.

The struggles don't end there. Every day on the campaign trail Trump's actions test the definition of normal. He calls for jailing his opponent. He openly admonishes sitting generals. He singles out minority groups for blanket condemnation. He goes after the spouses of his rivals. He questions the integrity of the election itself. He is endlessly hostile toward the media. All of this Trump does so often that it's a struggle to remember what's old news, by the standard of his behavior, and what is big news, by the standard of history.

But we're human, too. We can't help but become a bit jaded. There's a little game that's been going around, not a real one but a rhetorical one: name a campaign headline too crazy to be real. No one can do it. A few days ago, in the midst of such thoughts, BuzzFeed published a story with the headline: I HEARD IVANKA TRUMP TALK ABOUT "MULATTO COCKS."

When it landed I was working in an NBC trailer outside the Las Vegas debate site: the final face-off between Trump and Clinton. A few other correspondents and producers were there, and one of the senior producers had just stepped away for a few minutes, leaving his phone on the table.

"What did I miss?" he said, walking back in the room. For at least the next few minutes, *mulatto cocks* was our code for the sheer absurdity of 2016. Even more absurd, that Ivanka had to go on the record to deny it. Then that too was superseded, the unimaginable reduced to the forgettable and on again to the next thing.

Not tonight. Tonight is a remembrance for all the "mulatto cock" moments of 2016. There is a "telling you for the last time" quality to a lot of the stories, a "remember when" tone to many of the openings.

None of the stories are about sex or scandalous hookups. Jeremy and Ali are in love, which is not exactly scandalous. There are some more salacious rumors out there, of course, and a lot of duos whose relationships seem closer than producer and reporter, embed and advance staffer, Trump staffer and volunteer, or Trump staffer and Trump surrogate. Some of it could be true. None of it could be true. I don't know.

What I do know is that I haven't yet heard anything that comes close to one of the best lines in *The Boys on the*

Bus, about the extravagances of the reporters on the campaign trail in 1972. "These casual affairs," wrote Crouse, "produced at least three cases of the clap and one lawsuit—a stewardess, finding out on the last day of the campaign that her paramour was married, sued him for 'illegal acts committed over the state of Iowa.' "

Maybe we just aren't as wild, drunk, or fun as we used to be.

Is anyone?

But we still have some stories. One of my favorites comes from one of the producers here at this steak house tonight. Let's call him "Kenneth."

He recalls a little old lady who approached him outside of one of Trump's rallies. She wasn't wearing Trump gear and she seemed like sweetness personified.

"Excuse me," she said, walking right up to Kenneth. "What's your name? Are you part of the media?"

"Yes, I am, ma'am. My name is Kenneth."

She looked him right in the eye, holding his gaze.

"Fuck you, Kenneth," she said.

Another favorite comes from one of the correspondents, who remembers a call from one of Trump's comms people.

"Are we off the record?" the staffer asked.

"Sure," the reporter said. "What's going on?"

"Great. Off the record: Mr. Trump wants you to go fuck yourself."

The last story, at least the last story I can remember from that night, might as well be a metaphor for my life up to this moment.

A producer and a correspondent climb into a car after an early-morning live shot. They start the car, blast the heat, shift into reverse, and then promptly both fall sound asleep, driving backward across an empty parking lot. How do I know I'm not doing that right now?

There is no lakeside view for me at this Orlando "lakeside" hotel. It's October 25, the night after Bern's Steak House, the night before my thirty-third birthday. Anthony isn't at my hotel. He got the last room at the hotel inside the airport. Inside the airport! The jerk gets to sleep an extra thirty minutes, in what looks like a bright clean room, and roll right into the security line.

"See you at the gate!" he said a few minutes ago, waving good-bye. "Flight leaves at 6:20 A.M.!"

I am tired and irritable, looking out on a half-empty parking lot in a business zone a half mile from the Orlando airport. My jaw is clenched. My toes are digging into the carpet. I'm standing straight up, but my shoulders are hunched to my ears. The word *gnarled* comes to mind. I'm a gargoyle.

I try to shake it off, roll my shoulder blades back, stretch out my neck. But it's 10 P.M. In two hours I will be thirty-three years old, and I can't stop thinking about what I'd be doing if I were waking up tomorrow in London. I'd get a flat white at the hipster coffee shop around the corner and walk to work along the centuries-old cobblestones.

I'd file a *Today* show live shot or a *Nightly News* pitch, then I'd be off to lunch. A bowl of pho, perhaps with the ITV network's UK editor, Rohit Kachroo. We'd poke fun at our coworkers, gossip about who is trying to sleep with whom, and I'd have to *shhhh* him because he'd talk too loudly.

After some more scripting or a live shot, I'd grab a late-afternoon coffee with my cameraman Bredun Edwards. Over avocado toast, he'd tell me about the latest woman he is going to spend the rest of his life with. Then I'd tell him I want to go to Syria, or some other hot spot, and he'd tell me to love myself more. At the end of the day I'd get a pint or two with the rest of the bureau and I'd conspire to find a story that sends Chapman Bell or Lisa McNally and me to Rome or Marrakech. I really did have a story lined up for Tokyo, a feature piece about a "Jurassic Park" with animatronic dinosaurs. I'd be there filming right about now.

But instead I'm here, in Orlando, where I'll be up at 5:30 A.M. for another mad rush to another rally. I could set my alarm for 5:15 A.M. and take some of the stress off, but I'm still not a morning person. That extra fifteen minutes of sleep might as well be a brick of gold. Besides, I need it. I'm trying to hang on for two more weeks, until Election Day.

Don't ask me what comes next. I have no idea what I'm going to do or what I even want to do. I know I can't do another snowstorm. I can't be a domestic general assignment reporter again. I don't want to move to Washington, either. Would they let me cover politics from New York? Is London still a possibility?

I shut the curtains, put my hand on my stomach, and cringe. The queso dip I had for dinner is making me feel disgusting. Lately, I've been having a hard time getting into my clothes. The zipper won't close on my go-to pair of black pants, so I make Anthony safety-pin them together. Today the safety pin broke, too. It's bad. I avoid mirrors. I no longer recognize myself naked.

I simply don't exercise anymore—not even a plank or two in the hotel room. I say I'll watch what I eat, but then I eat whatever is in front of me. Usually that's something fried or processed or both. This Whole30 diet everyone talks about is intriguing. But you tell me what to eat in an airport at 6 A.M. that isn't bread, dairy, or sugar.

I get into bed at about 10:15 P.M. As usual, I'm exhausted but too wired to sleep. I write a pitch for the morning, knowing full well that it will be moot by the 8:45 A.M. editorial meeting. Trump will tweet at 6 or 7 A.M. or call in to *Fox & Friends* and say something so outrageous, he'll blow up our plans for a policy or voter story.

I am desperate to talk about his voters. I want to do a piece explaining that they don't care about the headlines. Either they aren't paying attention to them or they are discounting whatever they hear. The *Access Hollywood* tape, the women accusing Trump of sexual assault, the dark premonitions and lingering grudges—out here, in Trump's America, none of it is as big a deal as it is in New York or Washington.

I'd like to stop writing these pitches altogether, but I worry I won't get on TV. It feels like I need to remind my producers every day that I exist—that, yes, I will be covering Trump tomorrow. A lot of this is in my head. The fight for lead reporter isn't as rough as it used to be. The bosses are divvying up the assignments—trading off days on *Nightly News*. It's a logistical necessity, since MSNBC always needs a live shot and reporters are now hopscotching between rallies.

Besides, I'm enjoying MSNBC more lately, which is making things easier. I like the extended conversation, the way you can layer reporting and analysis, constantly pushing the story forward.

I finish my pitch and put down my phone.

I need cartoons. I need something completely different.

I turn on the TV and click around until I find Comedy Central.

Come on, South Park!

Nope. I keep clicking.

Who airs Family Guy?

TBS, TNT, USA . . . nothing.

Damn.

I shut off the TV and open my laptop.

BoJack Horseman, *I love you.*

Who wouldn't love a dark comedy about a washed-up, self-loathing humanoid horse actor, his feline agent, Princess Carolyn, and his best dog frenemy, Mr. Peanutbutter? It's brilliant. My favorite character is the incompetent cop: Officer Meow-Meow Fuzzyface. C'mon! That's comedy gold.

A packet of Justin's Honey Peanut Butter is poking out of the top of my backpack. I feel full and bloated, but the sticky mixture is nudging me on. I'm not moving, but somehow the packet is coming closer. The packet is moving right up to my mouth. My brain is saying no, but my hands are tearing open the foil. I knead the mixture and force it to ooze out of the opening in the corner like a Play-Doh snake. I bite down.

Kaboom!

The packet hangs from my mouth as I slump over like a junkie with a needle in her arm.

Tonight, BoJack is being framed for the murder of an orca dancer from a strip club called Whale World.

I told you this was a dark show.

Officer Meow-Meow Fuzzyface is on the case. I settle in and try to forget who I am and what I do for a moment. I make it ten minutes before I reach for my phone. There's a slew of rally logs, readouts, and planning notes for tomorrow's political coverage. *No, thank you.* I press my finger into the screen and drag it down to update my inbox. An e-mail pops up from Ben Jacobs at the *Guardian*. He's sharing his latest article. It has the kind of headline that will someday feel like a false memory.

Unbelievable.

JOE BIDEN'S WISH TO FIGHT DONALD TRUMP ECHOED BY CANDIDATE: "I'D LOVE THAT."

I don't particularly want to read another article, but this one is tickling my chin.

"Republican nominee Donald Trump implied on Tuesday that he would be willing to fight the sitting vice-president Joe Biden behind a barn."

It's even more bizarre than the talking horse on my laptop.

"The 70-year-old Republican nominee for president also labeled the vice-president 'Mr. Tough Guy' and said of beating up the 73-year-old Biden: 'Some things in life you could really love doing.' . . . The last time a sitting vice-president engaged in a major violent altercation with a fellow politician, Aaron Burr killed former secretary of treasury Alexander Hamilton in a duel on 11 July 1804 in Weehawken, New Jersey."

Heh.

"Love it," I e-mail back.

I want to put my phone back down and try to zone out, but it buzzes again.

It's Ben writing back.

"How are you? I am currently in a Red Lobster in suburban Tallahassee questioning a wide range of life decisions . . ." he says.

I'm never going to sleep. My body won't willingly put this phone down.

"I'm in bed watching BoJack Horseman and contemplating an Ambien," I write back. "I don't really want to be awake at midnight when I turn 33 in this d-list hotel by the Orlando airport."

I put my phone down and decide yes, an Ambien will help. I'd called my doctor for the prescription earlier today. I was all ready with my reasons: I can't relax. I shake my leg when I sit. I stare at my phone, endlessly refreshing Twitter and e-mail. I need some help beyond a glass of wine—which always leaves me feeling shaky in the morning.

It felt like he was waiting for the call. The words "I've been covering Trump" barely got out of my mouth before he said yes to the sleeping pills.

"Where's the nearest pharmacy?" he said.

I take a pill and set seven alarms on two phones for 5:30, 5:33, 5:35, 5:40, 5:45, 5:50, and 6 A.M. I plug in the phones. Turn out the lights. Leave the computer on. I like falling asleep to a show. I hope it will seep into my subconscious and stave off dreams about work. A few minutes later the room starts to blur. A man with a horse head is rocking me to sleep. A woman with a cat head and pearls is singing me a lullaby. I think an orca in a bikini is doing a striptease in the corner, but I could be hallucinating.

A current hums in the air, the voltage is up, and this rally is packed, just like every rally he's had since Friday. I'm standing on the riser, dead center, about a hundred feet from Trump himself. Although it's damn near high noon and the Miami sun is a killer, the crowd is on its feet, heads covered in Trump hats or Trump signs. Trump himself is onstage with a hundred-watt smile and his eyes locked on me.

I see a new Trump, or, more accurately, an old Trump. The version I haven't seen since the earliest days of the primary campaign. It's November 2, six days until Election Day. Just last week Trump was a dead man walking. He was a mopey, hard-luck candidate, down double digits in the polls, running out of money, energy, and time. At his rallies, instead of pushing ahead with a grand message, he would pause for a good wallow. The crowds still came to hear his greatest hits. But he would drift off into long tangents, nursing old grievances and dark conspiracy theories. He was full of blame and resentment, not vision and verve.

But now he's a man in a hot tub of pure adulation, pure possibility, splashing and smiling and enjoying the bubbles. "It's too bad those cameras aren't turning around to see this incredible group of people, with thousands more outside. Isn't it too bad? They don't do it," Trump says, and that's when I can tell something is coming. It's his usual riff on the media, his bull-pucky gambit about how the camera that's purposely fixed on him won't show his swooning crowds.

"But you know, if Hillary speaks and there's thirty people in a line, they say, 'Oh, the crowd is massive.' She has never had a crowd like this in her life," Trump continues. He's off the teleprompter now, wandering in the land of ad-libs.

"President Obama shouldn't be campaigning for her any longer—it's really a conflict, right? He ought to be working on jobs, on the border, on building up our military instead of campaigning for crooked Hillary Clinton. It's what he ought to be doing. He's in North Carolina, but we are going to North Carolina right after this, so—but we have two more today in Florida and we have massive crowds."

Trump often returns to crowd size, or television ratings, or magazine covers, any metric he can find to show he's richer, grander, more popular than thou. But in the last few days, the gap between Trump's boasts and the verifiable facts of his popularity has narrowed.

"There's something happening," Trump says, pointing to the crowd and then to me in the press riser. "They're not reporting it."

He's going to win," I said to Anthony. "It's over. He's going to win."

The moment when Trump's 2016 chances swung from almost hopeless to almost idiotproof was last Friday, while Anthony and I were on a flight into Cedar Rapids, Iowa. FBI director James Comey had just sent a letter to the heads of several committees in Congress. He wrote that he wished to "supplement" his prior testimony regarding the bureau's investigation into Hillary Clinton's private e-mail server.

This happened in the middle of the day, but I didn't hear about it until my flight landed ahead of a Trump rally that evening. There was no Wi-Fi on the plane, so the e-mails and other notifications arrived in a single avalanche of data. Just the words *Comey*, *Clinton*, and *e-mails* were enough to flutter my heart. This was a very big deal. I remember turning to Anthony in the terminal and predicting, right then and there, with all the confidence of a political neophyte: "He's going to win. Anthony, he's going to win. He. Is. Going. To. Win."

I believed it and I didn't; I knew it to be true and I was sure it was false. Anthony and I talked through the angles on the way to our live shot. The big-picture stuff was obvious and we knew the history by heart. Clinton's private server already captivated and enraged Trump supporters. It was the reason they began to chant "Lock her up!" in the first place, and it was the prime exhibit in Trump's case against her.

He called Clinton careless and unfit—in large part because as secretary of state, she had set up this private e-mail server. It had centralized her personal and professional e-mails, which she explained as a matter of convenience. After the *New York Times* revealed the existence of the server, however, back in 2015, it was clear that Clinton might have skirted the rules and potentially put classified information at risk.

From that flicker of misconduct her Republican opponents had started a major political blaze, and Trump had emerged as the arsonist in chief. Back in July, James Comey added to the heat at a press conference, where he called Clinton and her aides "extremely careless in their handling of very sensitive, highly classified information."

But at that same press conference he also acted as a fire-stopper for team Clinton, telling America that "no reasonable prosecutor" would bring a case against her. At least that's where the matter had rested. Trump could keep on railing against Clinton as a criminal and warning America against voting for a person who should have been indicted a long time ago. But he was at risk of sounding like a sore, old crank.

Until last Friday.

In Comey's letter to Capitol Hill, he said the FBI had found e-mails on the computer of Anthony Weiner, the estranged husband of Clinton's top aide Huma Abedin. Weiner is currently under investigation for sexting with an underage girl. The e-mails, according to Comey, appear to be "pertinent" to the Clinton server investigation, and the FBI will be reviewing them. It was a short, carefully worded letter about a narrow legal matter, but it roared through the political world like a rocket. At NBC News we learned of the letter at 12:54 P.M. Three minutes later the news was public.

"FBI Dir just informed me, 'The FBI has learned of the existence of emails that appear to be pertinent to the investigation,' " Jason Chaffetz, a Republican congressman from Utah, wrote on Twitter. "Case reopened."

"A great day in our campaign just got even better," Trump campaign manager Kellyanne Conway tweeted not long afterward. Trump, meanwhile, after so much dark talk of a rigged system, suddenly liked the system very much. He complimented Comey and the FBI, running with the news at an appearance in Manchester, New Hampshire, before coming to us in Cedar Rapids, Iowa.

"I need to open with a very critical breaking-news announcement," Trump told the Granite State crowd in a hotel ballroom. "This is bigger than Watergate."

He cheered the "courage" of the FBI, detailed his "great respect" for the bureau, and predicted that this move—righting the "horrible mistake" the FBI had made in closing the investigation—would "save their great reputation." The system, Trump said, with a self-deprecating lilt in his voice, "might not be as rigged as I thought, right? Right?"

The big question for Anthony (Terrell, not Weiner) and me, as we waited for Trump to arrive in Cedar Rapids, was how the news would affect voters. It would definitely give Trump back his biggest political hammer. The question was whether it would also bring hesitant Republicans to the polls and possibly keep hesitant Democrats at home. We'd spend the next few days pursuing an answer.

But first I had to do *Nightly* that night.

I've been doing *Nightly News* almost every night for a year, but it is still so stressful. The broadcast is only twenty-two minutes long, so each story doesn't get more than a minute and thirty seconds, a minute and fifty seconds tops. My scripts are never that short. I always have more to say. I always want to use more sound bites from Trump or more sound bites from voters. And then there are the days like last Friday, where it's not only stressful but nerve-racking.

It was just another report, yet I also knew it was one of those reports that could echo into the future, replayed in documentaries, archived to help future historians tell the story of 2016. I slipped in a pair of earbuds and stared toward the darkening horizon in Cedar Rapids. There was nothing there, but I knew there would be as my head filled with the sounds of a strumming guitar and a whispering melody.

"Much ado is all I see. And I feel like it's surrounding me. The crowd intrudes all day until I'm finally swept away."

The lyric repeated itself, drawing my eyes out farther . . .

"I'm finally swept awayyyyyyyy . . ."

And farther.

"I'm finally swept awayyyyyyyy . . ."

I felt a tap on my shoulder.

"Away . . ."

I closed my eyes.

Away . . .

I felt the tap again, harder.

"KATY!"

Fucking Anthony (again, Terrell, not Weiner).

"You're up in seven minutes," he said.

I took a deep breath, reached back to my collar to find my earpiece, the one funneling the *NBC Nightly News* feed, and checked in with the control room. I reached for my notepad and glanced through the tag I'd written for my story.

"You ready?"

Eric Marrapodi, the senior producer out of D.C.

"Yup."

"Your package is in. It looks good."

Five seconds of "tone" filled my left ear. Then beeps, one every second for ten seconds. A countdown to the show, then the violin and Lester's voice:

"Breaking news tonight, a bombshell from the FBI, eleven days before the election."

That night in Cedar Rapids the Trump campaign closed out the night with fireworks.

Real ones.

Now he's in Miami. The crowd is on fire—no extra explosives needed—and Trump is pointing a finger back to me on the press riser. "There's something happening. They're not reporting it," Trump says, and I look away for a moment. That's when I hear my name.

"Katy, you're not reporting it, Katy," Trump says. "But there's something happening, Katy. There's something happening, Katy."

He says it all with a smirk on his face, a lightness to his voice, and at the end he winks at me—I'm sure of it. He's being playful, a cad, but the crowd doesn't see any of this or they don't care. Everyone in the venue is booing me, and some try to get to my position. Security has to stop them.

I look back at Trump, lift a hand to my ear, and make a "Call me!" gesture. I want another interview, and I have enough experience now to know the crowds are less likely to stay angry if they see me playing along, taking it in stride, rather than stone-faced and subject to whatever they project on me.

But it doesn't seem to work.

Someone yells, "Katy sucks!"

I pull out my phone to write a reporting note to myself, typing "Katy sucks" into the subject line. Then I accidentally send it to the entire NBC News political e-mail list. On Twitter I can see the vitriol rising, the death threats blowing around like loose trash. I'm not surprised, but at the same time I'm sick of it.

The most aggressive members of the crowd are now as close as they can get, taunting me or, as my mother is no doubt worrying, thinking of doing much worse. A reporter from CNN sends our team a picture of a bald man who won't stop saying my name. "Watch out for him," she warns.

Right around then, my colleague Alex Seitz-Wald re-upped a little joke that rippled the last time Trump called me out, sending burbles of support under the hashtag "#ImwithTur." It's a play on Hillary Clinton's campaign slogan, "I'm with her," and it trends for much of the afternoon, long after the crowd clears out and a few lingering creeps slink home.

"Do me a favor," tweets the *Washington Post*'s Chris Cillizza. "Follow @KatyTurNBC to show her that doing your job—and doing it well—matters more than bullying."

I don't know how many followers jump on, but it doesn't matter.

Sometimes it's nice to have a notification that's not an insult.

The next morning it all feels like a dream, but there on MSNBC, Kellyanne Conway is talking about the incident.

"Well, he obviously didn't mean it in any malicious way," she says, which happens to be true, as well as irrelevant. The point isn't whether or not Trump is specifically interested in hurting me or any other journalist. It's that his comments put us in danger. Understanding this, one of *Morning Joe*'s hosts, Willie Geist, keeps up the questioning, trying to get Conway to comment on Trump's pattern. She pirouettes this way and that way, finally saying "I've spoken to NBC News about that issue, and I'll leave it there."

"Would you like to shed any light on it?" Willie follows up.

Yes, please shed light!

"No," says Conway. "Sorry."

But Willie isn't ready to leave it there, and I find myself

leaning toward the television in my hotel room. "Do you agree Donald Trump was wrong to call Katy out?" he asks.

Exactly! I'm dying to hear this answer.

"Well, I always prefer when he talks about institutions generally," Kellyanne responds, and then she says something that makes me doubt her moment-to-moment understanding of what's true and what's not.

"I had an exchange with Katy last night," she says.

No she did not!

Wait, I think. She's so calm. Is she right? Did Kellyanne Conway and I have an "exchange" last night and I just don't remember it?

I search my e-mail and my text messages, check my voice mail, and call Anthony. Nothing. Nothing. Nothing.

Finally, I check Twitter. There it is, in my direct message folder. It's not a note from Conway, not exactly, but there is at least a tweet she's decided to share with me privately.

It's a tweet from Keith Olbermann, the former MSNBC anchor I dated in my early twenties. Keith is a warm and gentle person who is also vicious when it comes to what he sees as BS. In the tweet Keith calls Conway a "fascist" and a "fucking idiot." This is a mystifying thing to send me. Is she trying to draw a parallel between her and Keith and me and Donald Trump? Is the point that us ladies are in some equivalent stew, whipped up by the men in our lives? That a presidential candidate's calling me out on national TV is somehow the same as a political commentator calling her a name on Twitter?

It's crazy, and yet at first I feel guilty, as if I have done something wrong because Keith wrote something nasty. It brings me back to the early days of my career when all any-one could find online about me was stuff about Keith, when I would hear whispers in the newsroom about how I supposedly got my job. It makes me feel small and worthless, which, I suddenly realize, is probably exactly what Kellyanne intended.

In a *Wall Street Journal*/NBC poll that comes out two days before the election, 62 percent of people say this presidential election has made them feel "less proud" of America. I'm at a Trump rally, a day later, the last day of the campaign, the second to last rally before the election, as vice presidential candidate Mike Pence warms up the crowd. By this point, the FBI has once again announced that there is no prosecutable crime in Hillary Clinton's use of a private e-mail server. But Trump has still been calling Clinton a risky bet, likely to be locked up.

The crowd in New Hampshire is frothing as Pence talks about Clinton. He's got a microphone, but in the middle of his speech another message cuts in, a maniac with a buzz saw of a voice, screaming out from the crowd. He's close enough to the press pen for me to hear but far enough from Pence that the God-fearing running mate keeps on talking as if nothing were happening. I don't know if Pence even hears this other man. Probably not. But I do, and I will never unhear him: not the man's message, and not the thousands of other voices that summarized 2016 by not shouting him down.

"Assassinate that bitch," the man said.

And the crowd said nothing.

"Assassinate that bitch."

And the crowd cheered on.

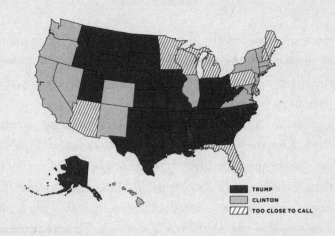

TRUMP VICTORY PARTY

NEW YORK HILTON MIDTOWN
1 A.M.–3 A.M., Election Day + 1

We don't know when Trump is coming out.

The election hasn't been called yet, not officially, but Trump is closing in on victory. He's won Florida and Ohio, the latter a state that has correctly picked the eventual president in every cycle since 1964. If he wins Pennsylvania, which now seems likely, he'll be at 264 votes in the Electoral College. He needs 270 to win, which means a win in Wisconsin, Michigan, or Arizona will put him over.

The room behind me is a madhouse. People who started drinking at 7 P.M., imagining an evening of sorrow, are now still going past midnight, joy in their voices. They're pumping fists, wiggling, spilling their liquor. For them, this is ecstasy. On the media side, this is usually the point in the night for stale champagne toasts to democracy and the pageantry of a free election.

But the shock is still too fresh, the surprise too thorough. Less than four hours ago, Chris Wallace of Fox News cautiously suggested that Trump might actually win this thing, then rapidly backpedaled. Less than twenty-four hours ago, *Politico* ushered in Election Day with a piece headlined TRUMP HOPES FOR A MIRACLE. To actually watch Trump's miracle come in is a shock like missing the last stair or sugaring your coffee with what proves to be salt. It's not just an intellectual experience. The whole body responds.

I hear Brian Williams and Rachel Maddow in my ear, then a toss coming from Brian: "Katy Tur is at Trump headquarters, having covered Trump for lo these five hundred and ten days." At the end there is no question, just my name and certain tone. I pick it up, not sure exactly what I am going to say but very sure of what needs to be said.

"I just want to remind people a little bit about Trump at this late hour and the questions that are still outstanding about him," I less say than hear myself say on live TV. "We still haven't seen Donald Trump's taxes. There are still questions about Donald Trump's business and how he would deal with a potential conflict of interest. He's talked about a blind trust going to his kids, but that technically is not really a blind trust. There are also still questions surrounding his relationship with Russia . . ."

On and on I go. I talk about his comments on a Muslim

ban, extreme vetting, the end of Obamacare, his cabinet, and more. In the middle of my report, I hear a new chant behind me.

"Call it! . . . Call it! . . . Call it!"

At about 1:40 A.M., Pennsylvania comes in for Trump, according to the Associated Press, pushing him to 264 votes in the Electoral College. NBC News hasn't called that state, and neither, for that matter, has Fox News, but the AP report jumps from one smartphone to another on the ballroom floor. He's one state away from a win. The crowd is in full gloat mode now. At one point, I see a man out on the ballroom floor, hollering, "Says who? Says who?" It seems like the ravings of a madman until I realize it's Michael Cohen, Trump's business lawyer, best known for an appearance on CNN back in August.

On that occasion, the anchor, Brianna Keilar, tried to establish as a premise, "You guys are down."

"Says who?" barked Cohen. "Says who?"

"Polls?" Brianna suggested. "Most of them. All of them?"

Five full seconds went by.

Then: "Says who?"

And so on.

Well, now he's still saying it, yelling it, in fact, and, by the looks of it, he's loving his life. He is, I will say, a vindicated man.

I'm in the mood for some vindication, too. I've been saying for months that Trump has more of a shot than the polls are giving him. I've been saying don't write him off. I've been saying it's not going to go as smoothly as the Clinton

campaign suggests. And at times I felt like a raving luna-
tic, the living embodiment of that old *Simpsons* meme: "Old
Man Yells at Cloud."

I get an e-mail from a producer who is shocked at the
results coming in. The subject line is: "Ummmmmm?"

The body of the e-mail is blank.

"Now I don't seem so crazy, do I," I respond.

"Get ready to move to D.C.," he jokes.

That's another theme of the night: the Welcome to
Washington jokes. In the bathroom, I see a drunk campaign
staffer, applying more makeup but willing to pause to welcome me to Washington. On the way out, I get an e-mail
from a colleague: "Welcome to Washington." "Welcome to
Washington," everyone says. But do I want to go to Washington? Do I have to go to Washington? My stomach flips
at the thought.

The night won't end. It's almost 2 A.M. and Hillary Clinton's campaign chairman John Podesta is expected to appear
any minute, perhaps to concede. The Trump ballroom is
rabid at the thought. They're chanting again. "Na-na-na-
na ... hey-hey-hey ... good-bye." But when Podesta appears,
he's in no mood to concede, telling supporters in New York:
"Let's get these votes counted and let's bring this home."

I call a senior source backstage to see where Trump is. He's
dying to come out, I'm told, but he's waiting for 270. At that
point the source breaks into a giddy Trump impression: "It's
tremendous. A movement. Believe me." The RNC claims it
nailed turnout. Trump started the movement. They knocked
on the doors. And their ground game brought it home.

At 2:30 A.M., Wisconsin falls for Trump, and now the ballroom is on its feet, chanting again, louder now: "Call it! Call it! Call it!" At about 2:40 A.M., the chants turn to cheers and then screams. Men are turning around and yelling at the press pen. But they aren't yelling at NBC or CNN, they're yelling at Fox News. Fox hasn't called it, not officially. The supporters think the media, even Fox, is trying to keep something from them. To refuse to let Trump win. But then—without warning—the ticker on the bottom of the Fox screens jumps past 270. Not even the anchors have been told yet.

"What is going on?" says Bret Baier, before finding his footing. "We are going to make this decision now: the Fox News decision desk has called Pennsylvania for Donald Trump. This means that Donald Trump will be the forty-fifth president of the United States."

Supporters are jumping, hugging, and yelling. They're waving bright pink WOMEN FOR TRUMP signs and red LATINOS FOR TRUMP signs. Stephen Baldwin, whose brother has been doing a ruthless impression of Trump on *Saturday Night Live*, is wearing a red MAKE AMERICA GREAT AGAIN hat and taking selfies with supporters. A couple is making out.

Baier promptly calls it "the most unreal, surreal election we have ever seen." I would call it unbelievable.

Ten minutes later, Trump is onstage, squared behind the podium, arms spread wide.

"Sorry to keep you waiting, folks," he says. "Complicated business."

Unbelievable.

EPILOGUE

Donald Trump went on to Washington, of course. He may or may not still be there by the time you read these words. There's a pretty good case that he didn't want to go in the first place; didn't like it when he got there; and wished for a way back into his old life that didn't involve scandal or shame.

In 2015, when he decided to run for president, he contacted *New York Times* reporter Maggie Haberman, offering her the exclusive news. She declined to break it—because she didn't believe it was real. As late as August 2016, senior GOP officials were reportedly preparing for the possibility of Trump dropping out, perhaps to launch his own Trump-branded cable channel.

Once Trump got to Washington, his own words and actions contributed to the idea that he might be a ripcord president, ready to parachute elsewhere at any moment. That is, once he actually got started with the work. He was inaugurated on a Friday, for example, but over the weekend he told a group of journalists that he didn't really consider Friday to be his first day.

"One of the first orders I'm gonna sign—day one—which I will consider to be Monday as opposed to Friday or Saturday. Right?" Trump said, according to *The Times* of London. "I mean my day one is gonna be Monday because I don't want to be signing and get it mixed up with lots of celebration."

Fair enough.

Once Trump got into the Oval Office, though, criticism of his work ethic followed. Trump's behavior made him an easy target. For one thing, he watched a staggering amount of TV news, including multiple morning shows, daytime cable, the Sunday shows, and the evening broadcasts. He called TiVo, the recording service for live TV, "one of the great inventions of all time," and he often Tweeted in tandem with whatever he had on.

Trump also got away from Washington regularly. Through his first six months in office, for instance, he was golfing at at least double the pace of his predecessor, and visiting his own resorts or clubs an average of once a week. If he was trying to gift himself more time for TV and golf, he did a decent job of it, handing the military more autonomy and failing to get around to appointing anyone to hundreds of government positions.

Still, part of him seemed to long for private life. In an interview to mark his first one hundred days in office, Trump turned wistful. "I loved my previous life," he said.

"I had so many things going." Then he admitted something usually left unsaid, something expressed in wrinkles, dark circles, and gray spots. "This is more work than in my previous life," Trump said. Yes, it's hard being president. People aren't usually surprised by that fact. Trump was. "I thought it would be easier," he said.

By the time the cherry blossoms popped in May of Trump's first year in office, the rumblings of an early exit had, if anything, intensified. "Does he want to be president?" wondered Fox News executive vice president and executive editor John Moody. "Might Trump feel that if and when he achieves his major goals—tighter borders, lower taxes, more American-made goods—he can declare victory and return to his successful career in the private sector?"

Trump had actually gotten some work done, of course. He rolled back hundreds of regulations—mostly through executive order. He got Neil Gorsuch onto the Supreme Court and got a version of his travel ban through the courts, at least temporarily. He also accelerated immigration arrests, which he had promised to do. And the economy hummed along.

But he hasn't gotten as much done as he has claimed and he has had far less popular support than a rating-watcher like him would have liked. His approval hovered near 40 percent through his first six months. Even voters in Trump counties want him to change. They may like his policies, but they don't like his style. They wish he would release his tax returns. They wish he didn't fire his FBI director, James Comey. They wish he'd stop fighting with the media. They wish he'd get off Twitter.

What can I say about Russia? I don't know. The country attacked America during the 2016 election, according to the U.S. Intelligence Community. They did it in part to help Trump become president. Multiple congressional investigations, an FBI investigation, and a special counsel are looking into whether the Trump campaign knew of this help, accepted it, and leveraged it in coordination with Russia.

The popular term is "collusion." The more legally problematic term is "obstruction"—the question of whether Trump has tried to block the investigation. The legal line has yet to be found. But when it is Trump may get his seemingly subconscious wish to leave Washington.

I, on the other hand, sometimes wonder if I should have gone to Washington. I didn't go myself. The lead reporter on a winning campaign typically goes to Washington as a White House correspondent. But I took myself out of consideration for the job. I still like being an outsider and, in my opinion, the most durable journalism is done by outsiders looking in.

Besides, I wanted a life. That friend who I crashed with when I got sick of hotels? He was a little more than a friend. By summer 2018, he'll be my husband. Or at least I hope he will be. Who knows if he'll still like me when he finally gets to spend some time with me.

I asked NBC to keep me in New York but still keep me on Trump. I cover him every day and for most of my waking hours, even if my MSNBC show only airs for one of those hours. I also still report and right now I can report that I don't know what happens next with Donald Trump. The second I do, I'll tell you.

ACKNOWLEDGMENTS

The process of writing this book was like nothing I've ever experienced before. But it was like just about everything else in my career in one important respect: I had help. My name is on the cover, just like my face is the one on the screen, but dozens of people have made that possible.

Jen Ortiz is a senior editor at *Marie Claire* who first invited me to write about my experience covering the Trump campaign. Over coffee at Trump Tower—where else—she decided that tweets and TV scripts had adequately prepared me to file a three-thousand-word glossy magazine story. I'll always be grateful for her lunacy. It's possible I would still have eventually turned my notes into some sort of published writing, but the team at *Marie Claire* launched my story with more style and more visibility than I could have hoped for.

Julia Cheiffetz is an executive editor at HarperCollins's Dey Street Books who saw a book in that essay and continued to see a book even after Trump won and I was too busy to write an outline, let alone a manuscript. I'm grateful to her for her vision, her persistence, her unshakable trust, her patience, and her thoroughly good taste. The book's title is hers; after two careful line edits, a lot of the book's grace is hers, too. I also want to thank the amazing team at Dey Street Books, including Lynn Grady, Sharyn Rosenblum, and Sean Newcott.

Of course, none of this would be possible without NBC News and MSNBC. Thanks to Andy Lack for seeing what I could do and putting me in a position to do it. Thanks also to Phil Griffin, who gave me room to run at MSNBC and caught me before I ran out of bounds. Deborah Turness demanded very early on that someone be assigned to cover Trump full-time. I'm thankful that she made that someone me. I'm equally indebted to Alex Wallace, Janelle Rodriguez, Sam Singal, and Don Nash for trusting one political neophyte to cover another. A special thanks to Noah Oppenheim for opening the next door.

I'm in awe of the whole *Nightly News* team, both in D.C. and New York. You make it look effortless even when I know for sure it is not. I owe pretty much all of you a bottomless glass of wine, but especially Eric Marrapodi, Jenn Suozzo, Talesha Reynolds, Mike Judge, and the other producers and editors. You stayed cool, even when I got hot.

Thank you to Ken Strickland, Doug Adams, Dave Forman, Freddie Tunnard, Vicky Blooston, and everyone on the D.C. Desk for solving the daily Rubik's Cube of travel and logistics and for somehow finding me a moment to breathe when I needed it most.

Our NBC crews are wizards. They made me look awake and put together when I was half-asleep and falling apart. Thanks in particular to Travis Rubury, Mike Ryan, and Larry Barr, who made a grimy old van parked on Fifth Avenue feel as comfortable as home.

The NBC News Politics team is a beast. My enduring gratitude to Mark Murray, Carrie Dann, Andrew Rafferty, Rob Rivas, and the army of folks behind the scenes who fact-checked, researched, and logged thousands of hours of political speeches and interviews. Emily Gold and Rachel Witkin deserve a special shout-out. You two are heroes. Not to mention Courtney McGee and Kenzi Abou-Sabe, who spent hours organizing and printing a foot-high stack of rally logs and reporting notes. Low five to Caitie Hawthorne for timing out who won what when on Election Night.

Thank you to David Verdi, Naomi Karam, and our security teams for keeping me safe. To Rich Greenberg, David McCormick, Ali Zelenko, and Mark Kornblau for allowing me to be me. To Omnika Thompson and Kerrie Wudyka and my entire MSNBC team for teaching me how to be an anchor while I was teaching myself how to write a book. To Aaron Freas for making the wish that brought me home. To Sharaf Mowjood for getting me started in politics (and keeping me out of a karaoke bar). To Alan Berger, Jonathan Lyons, and Ben Phelan for keeping me honest. To all my friends in London who cleaned my fridge and hung up my clothes.

I can't possibly list all my debts to Chuck Todd. So I'll go with a thank-you for his best piece of advice, which also happens to be his first piece of advice—the wisdom that launched me on this wild ride. When interviewing Donald Trump, ask short questions.

Thank you to my fellow Road Warriors: Kasie Hunt, Kristen Welker, Hallie Jackson, Jacob Soboroff, Kelly O'Donnell, Chris Jansing, Peter Alexander, Jacob Rascon, and Andrea "The Queen" Mitchell. I only worked as hard as I did to keep up with you all.

Thank you to my sources, you know who you are.

Thank you to Dafna Linzer for bearing witness; for writing a nomination letter to the Cronkite Awards that I am sure is the sole reason I won the award; and for only mildly scolding me for that *seven pieces of matzo* episode.

A big hug and a barrel of whiskey to the Traveling Trump press corps, including but not limited to: Tom Llamas, Major Garrett, Jim Acosta, John Roberts, Carl Cameron, Matty Hoye, Arden Farhi, Brandon Chase, Nick Kalman, Jennifer Girdon, Kristin Holmes, Ashley Parker, Jenna Johnson, Eli Stokols, Robert Costa, Phil Rucker, Jonathan Lemire, Jill Colvin, Evan Vucci, Eric Thayer, Carlo Allegri, Holly Bailey, Noah Gray, Sopan Deb (or is it Deb Sopan?), Ben Jacobs, Nick Corsaniti, Jose DelReal, Chris Snyder, Candace Smith, Ashley Killough, John Santucci, Jeremy Diamond, and especially Sara Murray.

I don't have the words to describe Erika Masonhall and Bradd Jaffy, but since they're two of my best friends, it doesn't matter. They hear from me enough. But Bradd, since I have you, I wouldn't be sitting here right now if you didn't volunteer me—out of nowhere—for that very first Trump story on *Nightly News* back in 2015. I will always be indebted to you for that. I also may never forgive you for it.

It's no surprise the best chapter in the whole book is about my incredible family, specifically my mom and dad. The Turs were never destined to be perfect. But perfect is boring. I love you, always.

To my arms and legs: Ali Vitali and Anthony Terrell. If I didn't write it down, I don't think anyone would ever believe us.

Finally, to Tony, do we have time to get married now?

ABOUT THE AUTHOR

KATY TUR is a correspondent for NBC News and an anchor for MSNBC. Tur is the recipient of a 2017 Walter Cronkite Award for Excellence in Journalism. She lives in New York City with her husband.